Place and the Scene of Literary Practice

The act of writing is intimately bound up with the flow and eddy of a writer's being-within-the-world; the everyday practices, encounters and networks of social life. Exploring the geographies of literary practice in the period 1840–1910, this book takes as its focus the work, or craft, of authorship, exploring novels not as objects awaiting interpretation, but as spatial processes of making meaning. As such, it is interested in literary creation not only as something that takes place - the situated nature of putting pen to paper - but simultaneously as a process that escapes such placing.

Arguing that writing is a process of *longue durée*, the book explores the influence of family and friends in the creative process, it draws attention to the role that travel and movement play in writing and it explores the wider commitments of authorial life, not as indicators of intertextuality, but as part of the creative process. In taking this 70-year period as its focus, this book moves beyond the traditional periodisations that have characterised literary studies, such as the Victorian or Edwardian novel, the nineteenth-century or early twentieth-century novel or Romanticism, social realism and modernism. It argues that the literary environment was not one of watershed moments; there were continuities between writers separated by several decades or writing in different centuries. At the same time, it draws attention to a 70-year period in which the value of literary work and culture were being contested and transformed.

Place and the Scene of Literary Practice will be key reading for those working in Human Geography, particularly Cultural and Historical Geography, Literary Studies and Literary History.

Angharad Saunders is a Senior Lecturer in Human Geography at the University of South Wales, UK. Her research interests revolve around the literary and cultural geographies of the late Victorian and Edwardian period. In particular, she is interested in the relationship between writing practice, as something more than a situated undertaking, and the imaginative worlds of the novel.

Studies in Historical Geography
Series Editor: Robert Mayhew

Historical geography has consistently been at the cutting edge of scholarship and research in human geography for the last fifty years. The first generation of its practitioners, led by Clifford Darby, Carl Sauer and Vidal de la Blache presented diligent archival studies of patterns of agriculture, industry and the region through time and space. Drawing on this work, but transcending it in terms of theoretical scope and substantive concerns, historical geography has long since developed into a highly interdisciplinary field seeking to fuse the study of space and time. In doing so, it provides new perspectives and insights into fundamental issues across both the humanities and social sciences. Having radically altered and expanded its conception of the theoretical underpinnings, data sources and styles of writing through which it can practice its craft over the past twenty years, historical geography is now a pluralistic, vibrant and interdisciplinary field of scholarship. In particular, two important trends can be discerned. First, there has been a major 'cultural turn' in historical geography which has led to a concern with representation as driving historical-geographical consciousness, leading scholars to a concern with text, interpretation and discourse rather than the more materialist concerns of their pred ecessors. Secondly, there has been a development of interdisciplinary scholarship, leading to fruitful dialogues with historians of science, art historians and literary scholars in particular which has revitalised the history of geographical thought as a realm of inquiry in historical geography. Studies in Historical Geography aims to provide a forum for the publication of scholarly work which encapsulates and furthers these developments. Aiming to attract an interdisciplinary and international authorship and audience, Studies in Historical Geography will publish theoretical, historiographical and substantive contributions meshing time, space and society.

For a full list of titles in this series, please visit https://www.routledge.com/Studies-in-Historical-Geography/book-series/ASHSER-1344

Place and the Scene of Literary Practice
Angharad Saunders

Elite Women and the Agricultural Landscape, 1700–1830
Briony McDonagh

Forthcoming:
Impure and Worldly Geography
Pierre Gourou and Tropicality
Gavin Bowd and Daniel Clayton

Place and the Scene of Literary Practice

Angharad Saunders

Routledge
Taylor & Francis Group
LONDON AND NEW YORK

First published 2018 by Routledge

2 Park Square, Milton Park, Abingdon, Oxfordshire OX14 4RN
52 Vanderbilt Avenue, New York, NY 10017

Routledge is an imprint of the Taylor & Francis Group, an informa business

First issued in paperback 2019

British Library Cataloguing-in-Publication Data
A catalogue record for this book is available from the British Library

Library of Congress Cataloging-in-Publication Data
Names: Saunders, Angharad, author.
Title: Place and the scene of literary practice / Angharad Saunders.
Description: Abingdon, Oxon ; New York, NY : Routledge, 2017. |
Series: Studies in historical geography | Includes bibliographical references.
Identifiers: LCCN 2016058269 | ISBN 9781472417640 (hardback) | ISBN 9781315600550 (ebook)
Subjects: LCSH: Authorship—Sociological aspects. | Geography and literature. | Place (Philosophy) in literature. | Authorship—Philosophy.
Classification: LCC PN151 .S28 2017 | DDC 808.02—dc23
LC record available at https://lccn.loc.gov/2016058269

ISBN: 978-1-4724-1764-0 (hbk)
ISBN: 978-0-367-24414-9 (pbk)

Typeset in Times New Roman
by Apex CoVantage, LLC

For my parents.

Contents

Figures

Acknowledgements

If this book began anywhere it began many years ago in a third year undergraduate module on the Geographies of British Literature. Under the guidance and subsequent supervision of Robert Mayhew, an undergraduate interest in literary geography became a postgraduate interest and over the intervening years has morphed into this book. Alongside my thanks to Robert Mayhew for inspiring my interest in literary geography, this book would not be what it is without the encouragement and friendship of Sheila Hones and the wider community of literary geographers. Over the years there have been many others who have shaped the literary practice of this book in some way, most notably Sarah Cant, Tim Cresswell and David Cooper. I would also like to thank Priscilla Corbett for her generous support in all aspects of the editing and production of this book.

My thanks must also go to Malcolm Thomas, Diana Wallace and Hefin Rowlands for helping to keep this book alive, and to Suzanne Jenkins and Ceri Price for taking on so much of the teaching that gave me the time and space to bring this book to completion. The book would not have been possible without the generous support of a University of South Wales Humanities Research Institute grant, a British Academy Small Research Grant and a Royal Geographical Society small grant.

Various libraries, archives and repositories have also been instrumental in the development of this book. Most notably Keele University Special Collections and Archives, the Potteries Museum and Art Gallery, the Bodleian Library, Princeton University Rare Books and Special Collections and the Rare Books and Manuscript Library at the University of Illinois, Urbana-Champaign. Several of these institutions gave permission for the use of the images and extracts that appear in this book, and for this I am most grateful. Thanks are also due to *Literary Geographies*, *Cultural Geographies* and the Taylor & Francis Group for permission to reprint work that has appeared elsewhere.

Finally, my thanks must go to Peter, Juliet, Rhonwen and Rhodri for so much stimulating conversation and support over the years, and to David, for while Trollope and t-cells don't always mix well, this book could not have happened without you. Finally, to Aneurin for keeping me company and keeping me focused in the last few months of this book's writing.

Prologue

In 1883 the publishing house of William Blackwood and Sons issued Anthony Trollope's *An Autobiography*. It was, for the time, a very unusual and controversial autobiography. It began benignly enough, echoing the conventions of the Victorian success story, with Trollope's reflections on his childhood, his time in the civil service, his move to Ireland and his desire to write. In many ways therefore the early parts of *An Autobiography* emulated the rags-to-riches style found in such fictional biographies as *Oliver Twist* (1838) and *David Copperfield* (1850). What followed, however, was quite different. It was a form that challenged the conventions of nineteenth-century autobiography; for as Robert Aguirre (2002) argues, it moved away from the popular tropes of autonomy, interiority, epiphany and conversion to address, very candidly, the relationship between the self and the practice of writing. In a chapter entitled, 'On Novels and the Art of Writing Them', Trollope offers his reader 'a few words on one or two points which my own practice has pointed out to me' (1992: 229). Trollope advises would-be writers to become intimate with their characters, to learn to live with them; he advocates self-criticism in order to achieve a more pellucid style; to avoid episodic narratives in preference for a fluid telling of the tale; and above all to practise constantly to fully understand and be able to render human nature. Later on the autobiography dwells on Trollope's writing practice, we learn of his custom 'to write with my watch before me, and to require from myself 250 words every quarter of an hour' (272), before calculating that this 'division of time allowed me to produce over ten pages of an ordinary novel volume a day, and if kept up through ten months, would have given as its results three novels of three volumes each in the year' (272). As the reader moves towards the closing pages of the autobiography they encounter a list of what Trollope calls his literary performances: every novel he had written alongside its year of publication and the amount received for it. These performances, 'I think are more in amount than the works of any other living English author . . . I have published much more than twice as much as Carlyle. I have also published considerably more than Voltaire' (362).

Although tinged with a certain arrogance and conceit this in itself was not what made *An Autobiography* so controversial. Rather, it was what such confidence revealed about the changing material practice of authorship in the late nineteenth century. This was a time of significant change within the British book world. The triple-decker novel was gradually being eclipsed in preference for single, cheaper volumes; there was a growing pressure for international copyright agreements to protect the creative imagination at home and abroad; literary agents were growing in number and this combined with the establishment of *The Society of Authors* in 1884, which aimed to protect and further the rights and interests of professional authors, meant that authorship was being slowly transformed into a commercial and professional undertaking (Law 2012). Running alongside this were more traditional conceptions of authorship that took much inspiration from Thomas Carlyle's argument that writers were part of the social aristocracy; they were public moralists charged with nurturing noble feelings and moral sensibilities among their readership (Collini 1991; Carey 1992). Consequently, authorship was not a career or profession; rather, it was a moral, social and spiritual calling. Thus, to suggest, as *An Autobiography* did, that it could be reduced to a set of mechanical instructions that could be taught and practised for economic gain was anathema to many within the book world.

An Autobiography is revealing for what it conveys about literary practice in the late nineteenth century. Although some were doubtful as to the quality and originality of Trollope's practice, with Henry James believing he 'abused his gift, overworked it, rode his horse too hard' (1883: 385), leading him to produce too much of the same, *An Autobiography* began to demystify what had, for a long time, been a relatively exclusive and hermetic calling. What is more, it began to suggest that a novel was far more than a material object: it had a history, it had come from somewhere and somebody. Thus, through *An Autobiography* we learn something of the scenes of writing: the social and spatial nature of a novel's making. We gain insight into where Trollope wrote, how he wrote, who was involved in his writing and how his geographical experiences, his travels, journeys and spatial imaginings were imbricated within his literary practice. Taking Anthony Trollope's writings as its starting point, it is to the scene of writing that this study attends. It is interested in literary practice not only as something that takes place – the situated nature of putting pen to paper – but simultaneously as a process that escapes such placing. Writing, it argues, is a process of *longue durée.* The act of writing is intimately bound up with the flow and eddy of a writer's being-within-the-world: the everyday practices, encounters and networks of social life. Such an approach opens up the influence of family and friends in the creative process, it draws attention to the role that travel and movement play in writing and it explores the wider commitments of authorial life, not as indicators of intertextuality, but as part of the creative process.

Although it begins with the work of Anthony Trollope this study is concerned, more broadly, with the period 1840–1910. In taking this 70-year period as its focus, this book moves beyond the traditional periodisations that have characterised literary studies, such as the Victorian or Edwardian novel, the nineteenth-century or early twentieth-century novel or Romanticism, social realism and modernism. Part of the reason for this is to argue that the literary environment was not one of watershed moments, and that there were continuities between writers separated by several decades or writing in different centuries. At the same time, it draws attention to a 70-year period in which the value of literary work and culture were being contested and transformed. As already noted, from around the mid-nineteenth century debates were emerging over the professionalisation and gentlemanliness of authorship, with concerns being expressed that what was once a 'calling', was now a material business (Altick 1957; Law 2012). Similarly, from the later nineteenth century through until at least the 1940s there were, John Carey (1992) observes, fissures arising over the relationship between the novelist, the novel and popular culture. Mass education, mass readership and mass production were challenging the nature, value and meaning of the novel and the novelist. Tensions arose, for instance, over the aesthetic function of the novel vis-a-vis its social purpose (Woolf 1959), over the democratisation of writing (Hilliard 2006) and over the intellectual quality and authority of the written word (Gissing 1974).

The period 1840–1910, therefore, played host to some very interesting changes within the British literary world. As such, it has been subject to quite considerable critical attention, but two texts in particular seem to anticipate this study most clearly. The first is Philip Waller's seminal work, *Writers, Readers and Reputations* (2006). Waller is motivated by an interest in how literary reputations, or the image and idea of the writer, were made and unmade during fifty years of social, cultural and technological change. It is very much the biography of a literary world; it reconstructs through the lives and works of countless writers and readers, the material, textual and imagined locations in which literary identities were forged and articulated. Whether this was the lecture theatre or the library, the pages of the newspaper or the periodical, or the establishment of a canon, it reveals how general changes in literary practice were intimately related to social change and social space. The second text is Christopher Hilliard's *To Exercise our Talents* (2006). This takes a narrower social and temporal frame, concerning itself with the democratisation of writing and the rise of the working-class author in the early twentieth century. What is particularly striking about this work is its focus on writing as a shared social practice. Correspondence courses, writing circles and self-help manuals helped nurture a writing world quite separate from the literary establishment in and of London. Taking inspiration from these two works, this study explores the interactions between the agency of the author and their social world in the making and re-making of the novel. Although it follows the broad temporal frame laid down by Waller, this study is more modest in its focus, taking not the literary world *per se*, but rather, a number of writers whose life and work, when taken together, spans this period. It examines the places in which these writers work, the way such placing shapes literary creation and the networks of movement and influence that informed the compositional process. In a sense, then, it is a finer grained analysis that takes not an era or a socio-economic group, but a number of individuals and attends to the epistemological relationships between their spaces and their writings or, more unfashionably perhaps, between their lived experiences and narration.

Despite what may seem an unfashionable leaning to some, geographers are keen to explore not just the thing itself but the relationship between things[1]. This inherently relational thinking has infused literary geography in recent years, most notably through Sheila Hones' (2008; 2014) theory of the spatial event of the text, which argues for the novel as a relationally generated event that is constantly made as it intersects with different social practices and geographical contexts. What this means is that the novel has no stability in space and time; rather, it happens – it is created, made and realised – where the different geographies of author, text and reader encounter and interact with one another. The power of this theory, as Hones observes, is that it is able to accommodate differences in approach across the field of literary geography, for it regards each approach as a distinct spatial event. Despite the inherent flexibility of the 'spatial event of the text', and its recognition that creation occurs across the event of the text, our geographical understanding of it remains a little skewed, for it has tended to follow the thinking of reader-response criticism, where emphasis is placed on the agency of the reader and their interaction with the text as a finished product. Little attention has been directed to the 'other side' – to the intersection and distribution of social agencies that occur in, or are pivotal to, the creative process and the making of a text as a material entity. It is timely, therefore, to return to the world before the text and to resuscitate the writers who reside therein, for this is a lived world we know little about and it is not one that is wholly separate to the imagined world of the text.

It is time then, to say something more about the three writers who live at the heart of this study. Anthony Trollope (1815–1882) we have already met, but alongside him are John Galsworthy (1866–1933) and Arnold Bennett (1867–1931). At first sight one might ask why these writers, for none is the name of their age, few commonalities suggest themselves and temporally they are a trio that seem a little unbalanced, skewed towards the early twentieth century. If Trollope is eclipsed by Charles Dickens and George Eliot, we find Galsworthy and Bennett hiding behind Joseph Conrad and Ford Madox Ford[2]. They are all overshadowed, within the popular imagination at least, by more iconic and experimental wordsmiths. This, however, is partly their draw. Trollope is often seen as a conventional, even mundane novelist; his measured, easy narrative tone, his aversion to analysis and introspection, and Henry James' belief that he 'has never troubled his head nor clogged his pen with theories about the nature of his business' (1883: 386), make for an undemanding reading experience (Williams 1974). Meanwhile, interest in Galsworthy and Bennett is often dwarfed by modernism's 'gigantic shadows' (Ardis 2002: 4), which have twisted and distorted the cultural field of the early twentieth century so that little can or is said about a novel unless it is expressed in terms of modernism (Hapgood 2000a). It could be suggested then, that they are all outsiders or have been constructed as occupying a position peripheral to the inner circle of Britain's literary canon. Yet, as those working against these popular periodisations and evaluations of literary history argue, each of these writers was engaged in challenging and transforming the literary structures of their time (Ardis 2002; Hapgood 2000b; Skilton 1994). Although following in the spirit of these scholars, this book is interested not in decoding the challenge contained within these writers' works *per se*, but of their thought-in-action, or, put another way, it is less concerned with what a text means and more concerned with a text's creation (Hawkins 2013). It is here, in the turn to creation, that further commonalities between these three writers emerge.

Although retrospectively valued as having less skill than some of their contemporaries what is interesting about Trollope, Galsworthy and Bennett is that each, to varying extents, took a reflective interest in their own practice; for instance, Anthony Trollope was a prolific writer and one often considered to have taken writing to an industrial level. Trollope would set himself daily targets and adopt particular habits and methods of writing to achieve these, and unlike many of his contemporaries he was very public about his writing practice. Arnold Bennett was in some ways quite similar, for he kept regular records that were very explicit about the relationship between daily routine and daily output. Together these two writers cast a light on the everydayness of writing and on the way it is bound up with being and moving within the world. Moreover, their concern with movement, routine and skill suggest that creative labour is not only a mental happening but something that is bound up with their being-within-the-world. John Galsworthy, by contrast, was far more reticent about the everyday details of his writing, but what was startling about his literary practice was its sociality. Galsworthy worked through his friends and family and they became central components, as critics and commentators, in his compositional process. His working practices then, were evocative of what has been termed 'distributed creativity' (Hawkins 2013: 10): a recognition of the others, the time and spaces involved in literary making. Thus, what links these writers is an interest in practice; an interest in the machinations of making and in expressing, albeit in very different ways, what went on along this process of making. By drawing our attention to the peoples and places implicit within the literary practice of these writers, this book provides a way of understanding literary culture as something more than what occurs within metropolitan centres, as something more than the world of editing, publishing and reviewing and as something that happens between as much as within any one individual. Trollope, Galsworthy and Bennett give us access to some of the hidden corners of literary culture in the period 1840–1910.

Structure of the book

The first part of this book explores the relationship between literature and geography. What follows is structured around three key themes. The first is place and it picks up on debates ongoing in histories of science that argue that where science happens, where it takes place matters to its making and meaning. It begins inside Les Néfliers, Arnold Bennett's house in France and explores the interior geography of this house (along with that of some of his earlier residences): the design, the décor and the manner in which it was assembled and contested and what all this meant to Bennett's writing practice. It then moves outwards into John Galsworthy's neighbourhood of Holland Park and examines this space as what Nigel Thrift (1996) terms, a productive locale; as something more than the backdrop to Galsworthy's literary practice. Drawing on the work of Doreen Massey and Sara Blair Chapter 2 explores this place not as a settled and coherent entity but as something living, lively and articulate, which makes its presence felt in Galsworthy's fiction. Finally, this section alights on Trollope's Ireland to address the relationship between the place of writing, reading and representation. Drawing on Trollope's choice of Irish subject matter for his first novel, this chapter argues that audience expectation was a critical force in the creative process. What links these three chapters is an interest in the different scales of literary production: the local, the regional and the national and how these scales play into the literary craft in very real ways.

Part II engages more directly with the idea of the scene, doing so through a focus on the spatialisation of writing routines. Once more it takes the three writers individually and begins with Arnold Bennett and the idea of his writingscape. The writingscape is an elaboration of Tim Ingold's (1993) notion of the taskscape, the array of tasks through which our everyday lives are produced and lived, and is used to turn our attention to the way literary craft always spills beyond the confines of a study or writing room. As Chapter 4 shows, crafting fiction was part of an author's being and moving-within-the-world. The next chapter turns to Anthony Trollope's writing journals. It draws upon work in textual geography that explores the spatiality of the textual form but brings this thinking to bear not on the finished text, but on the draft or preparatory materials that were central to Trollope's writing practice. The chapter argues that the spatial layout of these materials was central to the production of meaning both within the creative process and in the text under construction. The final chapter in Part II approaches routine from a somewhat different perspective. Taking what is often considered Galsworthy's most autobiographical novel, *The Man of Property*, it looks at the process of translation from the lived to the imagined world. This is a relationship that has generated some unease within academia for the way it encourages scholars to seek meaning about the latter in the former, when in fact the relationship between the two is far from contiguous. Yet far from suggesting verisimilitude between one world and the next, this chapter, taking inspiration from Catherine Brace and Adeline Johns-Putra (2010), probes the process of representation, arguing that lived experience is often central to creative labour or at least the creative spark, and explores how this lived experience is changed and transformed through the process of narrativisation.

The final part in this book develops the idea of *en-route* writing. Borrowing again from histories of science that are interested in the route to print, or put another way, in the way experience travels and transforms from the lived world to the textual world, *en-route* captures the process of mediation and negotiation – of world, of word, of content and of tone – that is part of the practice of writing. To date, *en-route* is an idea that has not been played into the literary world, partly because of imaginative literature's very imaginativeness and the disjuncture this has with the real world, but also because of the looser, more fluid and more fragmented way in which aspects of the real world fall into the imagined world. As the chapters in Part III argue though, this does not discount us from playing together the literary with the idea of *en-route*, for there is always a route to writing. Within the literary world, this route might not come from extensive travel that shapes form and content, but as Charles Withers observes, being *en-route* is as much about the practices, techniques and technologies of writing and these often come from the world and shape the imagined world. Moreover, as he claims, correspondence is an 'epistemic desideratum' for 'what is written about should correspond in some way to the world thus described' (2011: 85). Withers is obviously concerned here with the route to geographical knowledge, but even in literary circles there is a recognition, sometimes grudgingly made, that there is a correspondence between the real world and the imagined world (Booth 1961; Genette 1980; James 2008). These ideas inform both the two chapters in this part and the epilogue. In the first, we join Trollope on his travels and explore how the act of being-on-the-move shaped the imagined worlds he created. Drawing on his travel diaries, it explores how Trollope's own mobility was mediated, condensed and interpolated into his fiction. Chapter 8 turns its attention from the writer *en-route* to the movement and circulation of the things and objects of their world. Hence, it attends to Galsworthy's correspondence, to the letters that circulated between him and Joseph Conrad and Edward Garnett. This correspondence, it argues, transported and transformed the imagined world of Galsworthy's fiction; it reveals the way interpretation happened and meaning was construed differently across space as different readers evaluated the authenticity, integrity and consistency of the world Galsworthy had created. The final chapter, and epilogue to the whole, weaves together ideas of *en-route*, with the book's overriding interest in the place and the scene of literary practice. It takes the humble postcard, a papery object that circulates and travels, but which, unlike the letter, has a more constricted spatial form that makes its connection to literary practice more ambiguous, and suggests how the postcard gives us insight into both the scalar nature of writing's happening and the different ways it unfolds over space and time. The focus, then, is not so much on how writing travels, but how the form travels, spatially and temporally, and what this can tell us, more broadly, about the on-goingness of literary practice.

Notes

1 This is in reference to the predominance of close-reading strategies within literary studies, with studies of authors often seen as reducing textual analysis to biography. More will be said about this in subsequent chapters.

2 Ford Madox Ford was known as Ford Madox Heuffer until 1919, when he changed his name by deed poll. Here and throughout Ford Madox Ford will be used for this is the name by which he is most commonly known today. The exception to this is in contemporary quotes that directly refer to Ford Madox Heuffer.

References

Aguirre, R. 2002. Cold print: professing authorship in Anthony Trollope's autobiography. *Biography*, 25 (4), pp.569–592.

Altick, R.D. 1957. *The English Common Reader: A Social History of the Mass Reading Public 1800–1900*. Chicago: Chicago University Press.

Ardis, A. 2002. *Modernism and Cultural Conflict, 1880–1922*. Cambridge: Cambridge University Press.

Booth, W. 1961. *The Rhetoric of Fiction*. London: University of Chicago Press.

Brace, C. and Johns-Putra, A. 2010. Recovering inspiration in the spaces of creative writing. *Transactions of the Institute of British Geographers*, 35 (3), pp.399–413.

Carey, J. 1992. *The Intellectuals and the Masses: Pride and Prejudice among the Literary Intelligensia, 1880–1939*. London: Faber and Faber.

Collini, S. 1991. *Public Moralists: Political Thought and Intellectual Life in Britain, 1850–1930*. Oxford: Oxford University Press.

Dickens, C. [1850] 1992. *David Copperfield.* Ware: Wordsworth Editions.

Dickens, C. [1838] 1994. *Oliver Twist*. London: Penguin.

Genette, G. 1980. *Narrative Discourse: An Essay in Method.* New York: Cornell University Press. Trans. J. Lewin.

Gissing, G. 1974. *Demos: A Story of English Socialism*. London: Prentice Hall.

Hawkins, H. 2013. *For Creative Geographies: Geography, Visual Arts and the Making of Worlds*. London: Routledge.

Hapgood, L. 2000a. Transforming the Victorian. In L. Hapgood and N. Paxton, eds. *Outside Modernism*. Basingstoke: Macmillan, pp.22–42.

Hapgood, L. 2000b. The unwritten suburbs: defining spaces in John Galsworthy's The Man of Property. In L. Hapgood and N. Paxton, eds. *Outside Modernism*. Basingstoke: Macmillan, pp.162–179.

Hilliard, C. 2006. *To Exercise our Talents: The Democratisation of Writing in Britain.* London: Harvard University Press.

Hones, S. 2014. *Literary Geographies: Narrative Space in Let the Great World Spin*. New York: Palgave Macmillan.

Hones, S. 2008. Text as it happens: literary geography. *Geography Compass*, 2 (5), pp.1301–1317.

Ingold, T. 1993. The temporality of the landscape. *World Archaeology*, 25 (2), pp.152–74.

James, D. 2008. *Contemporary British Fiction and the Artistry of Space: Style, Landscape, Perception.* London: Continuum.

James, H. 1883. Anthony Trollope. *The Century Magazine*, 36 (July), pp.384–394.

Law, G. 2012. The professionalization of authorship. In J. Kucich and J.B. Taylor, eds. *The Oxford History of the Novel in English, Vol. III: The Nineteenth-Century Novel, 1820–1880*. Oxford: Oxford University Press, pp.37–55.

Skilton, D. 1994. Introduction. *The Bertrams*. By Anthony Trollope. London: The Folio Society.

Thrift, N. 1996. *Spatial Formations*. London: Sage.

Trollope, A. [1883] 1992. *An Autobiography*. Oxford: Oxford University Press.

Waller, P. 2006. *Writers, Readers, and Reputations: Literary Life in Britain 1870–1918*. Oxford: Oxford University Press.

Williams, R. 1974. *The English Novel From Dickens to Lawrence*. St Albans: Paladin.

Withers, C. 2011. Travel, *en route* writing and the problem of correspondence. In M. Hvattum, J.K. Larsen, B. Brenna and B. Elvebakk, eds. *Routes, Roads and Landscapes*. London: Ashgate, pp.85–98.

Woolf, V. [1923] 1959. Mr Bennett and Mrs Brown. In *The Captain's Death Bed and other Essays*, New York: Harcourt Brace Janovich.

Introduction

The place and the scene

Scene: theory, history and method

What do we mean by the *scene* of literary practice? Scene is a word rich in meaning. Etymologically it derives from the Greek work skene (σκηνή), which literally means tent or covered place. There is some suggestion that in ancient Greece skene was a word with artistic associations, being a structure at the back of the stage that facilitated the entrance and exit of actors (Taplin 2002) others, however, argue that it was merely a generic designation of place rather than something innately theatrical (Shusterman 2001). It is the Latin word scaena, meaning the background before which performances were given, which offers a clearer alignment of scene with place: scene is the arranged, designed and observable setting of theatre. Scene has never really lost this latter connection with dramatic space, yet today its meanings are more numerous. If we turn to dictionary definitions it has as many as 15 different meanings, which include a landscape, picture or place of action; a view or spectacle; a stage performance; a dramatic incident or episode; the entry or exit of actors and the uncomfortable display of feelings (The Chambers Dictionary 1998). It is, then, understood as something more than a backdrop to action and this multiplication of meanings can, in some respects, be traced to changes in dramatic art that occurred during the nineteenth century.

Over the course of the nineteenth century a number of French writers were lamenting the very woodenness of dramatic art both on the stage and in literature. There was a sense that dramatic art made its mimesis too evident; it was too obviously a representation rather than a seamless extension of life (Freeborn 1960; Lehan 2005). Émile Zola for instance, was critical of acting's very rigid conventions, such as its preference for overly-formal entrances and exits, the quite gauche interaction between actors and the seeming disconnection between actors and the scenic backdrop, which all made acting so obviously acting (Pavis 2012). It is this point of disconnection that is most important here for it suggests that scene was gradually coming to encompass more than mere backdrop; actors were part of not separate to the scene. In France this critique was bound up with the emergence of *mis-en-scène*: the professional art of dramatisation. *Mis-en-scène* was the spatial and temporal ordering of a performance, and it began to remove drama from the whim of the actor and insist upon the tight and holistic control of each dramatic scene. This did two, slightly paradoxical things. Firstly, and as Richard Shusterman (2001) observes, it led to the very clear framing of a scene. A scene became the sum of its spatial and temporal properties – lighting, stage directions, backdrop, props and people – which together created both an intense and dramatic happening and one that was discrete and delineable. Secondly, and somewhat conversely, it unframed the scene. In using the analogy of a frame and framing, scene can be likened to a picture: it is a composition that holds time and motion in place, it insists upon a particular view and view point and it has a logical order. Yet as Robert Edmond Jones (1941) argues, a scene should not be a picture made for our visual consumption, but rather, an image in which we, as the audience, are a part. A scene, therefore, becomes not merely an observable object, a backdrop to action, but a lived practice that engages and affects us.

There are similarities here with Ludwig Wittgenstein's (1973) picture-theory, which argues that one of the main functions of language is to allow us to picture things within our mind. Yet pictures work to hold us captive, for they belong not to us, but to others: they are pre-imagined. We are not, then, the author of a picture, but we come to rely upon it as a medium of our consciousness, allowing it to efface our imagination. Thus, our understanding of the world comes through pictures not experience; pictures present us with a world and they tell us about that world, but they do not immerse us in it or enrol us in its making. Our imagination, in contrast, is in constant dialogue with the world. It is a dialogue that allows us to imagine other possibilities present within an event or happening (Shotter 1993). A picture is the creature of another: it is conveyed to us and it is not ours. The imagination is creative, it makes meaning and it belongs to us. There are recognisable problems with Wittgenstein's picture theory, not least that any picture is, in essence, our own creation. A picture may be signified by others, but it is only we who can draw, fill and colour a picture in our mind. This aside, the broader significance of Wittgenstein's idea lies in what it suggests about ways of knowing. The intimation is that a picture tells, it gives us 'knowledge', whereas an image shows and reveals knowledge in the making.

We find parallels within literature, where Percy Lubbock (1921) draws a distinction between pictorial (sometimes referred to as panoramic or summary) and scenic methods of narration. Lubbock takes W.M. Thackeray as indicative of the pictorial method, for 'in his story we need him all the time and can never forget him. He it is who must assemble and arrange his large chronicle, piecing it together out of his experience' (1921: 113–114). The pictorial method is one of telling, for the story is communicated to us through the voice of the author or narrator as they condense, mediate and reflect the action to the reader. In comparison, the scenic method is one of showing in which the voice of the author is absent or obscured. It is a method that immerses the reader, through the use of dialogue, very vivid description or, by the time we reach modernist fiction, interior monologue, within the immediacy of the narrative's dramatic happenings. Where the pictorial method reports events, or gives us knowledge, the scenic method involves us in its duration and development; we are immersed within the passage of time and the flow of space and we make meaning as events unfurl and happen. Although Lubbock (1921) believed that all literature should aspire to showing rather than telling, few novels operate solely through the scenic method. Instead, most interweave scenic and pictorial methods of narration: strong periods of action that convey only minutes of lived experience are interspersed with broad strokes that may cover many years. This interspersing of methods gives rise to discrete scenes – a series of spatial and temporal happenings. From this it is easy to construe the scene as a moment of dramatic action – it is where the singularity and richness of life happens. Yet, we must not be misled into conflating scene with happenings that are big and exceptional to the exclusion of those that are small and seemingly inconsequential. As Gérard Genette (1980) infers, scenes are not merely big spatial and temporal happenings, but small ones too; they can immerse us in mundane social and psychological worlds, in which not much happens. A scene, therefore, can almost be a slowing of action, a focusing or gathering in, upon a highly particularised time and space.

Scene, then, is both a method of writing the world in duration and a focalisation upon a particular time-space. It is this duality that makes it so useful when thinking about literary practice. The turn to practice within literary studies recognises that a text is not merely a finished product awaiting interpretation, but is equally a real and imagined entity that is made within the world (Bushell 2009; Saunders 2008; 2010: Brace and Johns-Putra 2010). It takes inspiration from practice theories that seek to complicate objective and subjective understandings of the world, by taking the body as our immediate nexus with the world. Martin Heidegger (1978), for instance, argues that we know not merely through spectatorship or contemplation, but through being busy in the world. Busyness does not denote physical exertion, but rather, concerned engagement. In moving, working and socialising within the world, the world enfolds us and affects us. Concerned engagement is not though, a necessarily conscious activity. It is, what Maurice Merleau-Ponty (2013) might term, pre-reflective; it exists before thought. Our practices can be intuitive, routine, ceaseless and often unthinking, but it is through and within them that we get caught up in the flow and eddy of the world. In urging us to understand the active processes and modes of being through which our material world operates, practice theories encourage literary studies to move beyond the meaning in the text, to examine the making of this meaning. Thus, rather than being confined by the parameters of the printed wor(l)d, literary scholars are increasingly interested in the 'making' of this wor(l)d; the manner in which it unfolds and is crafted over time and space.

Scene, therefore, becomes a useful way of capturing the unfolding of literary practice, for it draws our attention to both the place and process of this unfolding. If we follow Genette (1980) and think of scene as a slowing and focusing of action it enables us to think about the taking place of literary practice. In his exploration of scientific practice, David Livingstone (2003) observes how practices are conducted in place and bear the traces of these places in numerous ways. Place dictates certain forms of behaviour, deportment and social interaction; it controls access and movement; it has its own vernacular and it permits and evaluates knowledge in particular ways. Although the importance of place to writing is acknowledged, we know, for instance, that Dylan Thomas chose to write much of his work in his garden shed and in *A Room of One's Own,* Virginia Woolf muses on a writer's need for a personal and private space, both literal and figurative, for creativity, such examples are more anecdotal than empirical. Modernist studies have done most perhaps to explore the relationship between place and writing in more rigorous ways. Where Morag Shiach (2005) examines the domestic interiors of various writers as an alternative way of reading the modern city, Andrew Thacker (2005) explores London, and Sara Blair (2004) the more specific locale of Bloomsbury, as places productive of particular styles and forms of writing. In creation studies, Charlotte Doyle (1998) discusses the writingrealm as a designated writing space in which writers adopt particular ways of being and doing. Although these studies are all actuated by place, it tends to be the idea or tenor of place, not the particularities or generative properties of place that are important: the significance of place lies in its physical, cognitive or cultural separateness, not its lived materiality. Thus, place matters to literary writing, but this significance is worthy of greater refinement in terms of both its materiality and lived sociality.

Place, though, presents problems for any consideration of literary practice, 'for locating writing in place presumes that it is a stable, containable practice' (Saunders 2010: 443). Practice by its very nature suggests duration, and while this can be duration in place, the putting of pen to paper, or finger to keyboard, this is only one expression or dimension of literary practice. Literary practice is so much more than the physical act of writing. A sense of this is captured in Sally Bushell's (2009) idea of 'text-as-process', which urges us to look not at the material text, but at the compositional materials that come before the written wor(l)d: the rough drafts, the revisions and the manuscripts that go into its making. It is an approach that begins to register the *longue durée* of writing, but literary practice, this book argues, is more than an attention to a text's compositional materials, it is an attention to the compositional nature of lived experience. Put another way, and taking inspiration from Nedra Reynolds (2004), it is understanding composition as a spatial practice. It takes form through the ways in which we dwell, move through, inhabit and cohabit in the world: our daily routines, social encounters, journeys and travels. Arnold Bennett, for instance, frequently noted the role that walking played in his creative process, while Anthony Trollope was known for his love of writing whilst travelling, often choosing to write on trains and boats. Within both of these examples writing is part of a larger experience of being-within-the-world; while the writers in question may well sit down to write in particular places, their writing simultaneously exceeds and is made beyond and between these places. This is what Brace and Johns-Putra (2010) term the beings of authorship: the life that exceeds the physical act of writing, but which is never separate from it, for it ineffably inspires, influences, and stimulates the creative process. In short, literary practice is part of the ceaseless unfolding, the relational happenings, and iterative nature of the everyday. Thus, this is where the other significance of scene matters – scene as method – for as a way of writing that seeks to reveal the world *in media res*, it offers us a conceptual device, or way of enfolding ourselves within, the unfurling of creation across time and space.

The scene of writing

From early studies that privileged the author as the repository of meaning and sought to evaluate the representational accuracy of realist literary spaces, literary geography in the 1990s began to develop a more critical dialogue with literary studies. Taking inspiration from New Criticism and post-structuralism, Marc Brosseau (1994) urged literary geographers to move away from a focus on the agent of the text towards the agency of the text; to attend not to what a text says about space or author, but to how it says it. Although this work seemingly displaced the author as a subject of analysis it led, through its emphasis on close reading, to a more incisive interrogation of agency. A focus on the agency of the text, for instance, was joined by a growing awareness of the agency of the reader, with James Kneale (1999) exploring the co-production of textual meaning through the interplay of authorial conventions and reader expectations. Sheila Hones (2008; 2014) meanwhile, in developing a literary geography motivated by spatial theory rather than by literary theory, argues for a more relational understanding of agency. This positions the text as a spatial event that is brought into being by the intersection of multiple actors and agencies – authors, texts, readers – across time and space. It is a theory that suggests that all texts are performances, which coalesce through the interactions of the multiple spatialities of their different agents. As such, it becomes difficult to locate meaning within particular agents, for the text has no stability over time and space. Instead, meaning emerges differently with each happening of the text.

Thus, literary geography has begun to develop more sophisticated and spatially nuanced ideas about the nature and location of agency, yet there is scope to develop these ideas further by engaging in dialogue not only with literary studies but with English studies more broadly. To date, literary geographers have followed the predominant theoretical positions of literary studies that are premised upon a denial of origins; on a concern with the production of meaning and not on the happening of the creative process (Bushell 2009; Dawson 2005). Consequently, literary geography has tended to follow the thinking of reader-response criticism, where emphasis is placed on the agency of the reader and their interaction with the text as a finished, material product; it attends to the reading practices of individuals, editors and critics as they engage with the novel as a published form (Hones 2008; Noxolo and Preziuso 2013; Yapp 2011). What this misses, however, is those practices that precede a text; its bringing into being, or happening, as a material entity. This is the 'other side' of the spatial event of the text and to develop a richer understanding of the novel as a geographical phenomenon we need to look not just to how it happens through the spatial relationships of its reading, but to how it happens through the spatial relationships of its writing.

English studies is broadly divisible into literary and textual studies. Where the former is concerned with what Jerome McGann (1991) terms the scene of reading – in what one does with a text – the latter is interested in the scene of writing – with how texts come in to being. The scene of writing is a rich field, encompassing questions of editorial practice and authorial intention (McGann 1991; Parker 1983) compositional strategies (Dawson 2005; Lunsford and Ede 2012; Reynolds 2004), attribution (Love 2002) and influence (Farrell 2001; Glyer 2007; LeFevre 1987), but where many of these threads come together is in their growing recognition of writing as an outcome of social relationships, or, put another way, as a social practice in which authority is distributed between numerous agents, or, as Derek Attridge (2004) terms them, others. Early work sought to elucidate the role that others played within textual composition, with McGann (1991) exploring the material traces that editors and publishers left within a text or its variants, while Reynolds (1993) was keen to reveal the gender identities that were hidden or registered within a text. It was not so much the social nature of writing that was revealed in these works though; rather it was the social nature of the text. A more enlivened sociality is evident in influence studies, where emphasis is given to the role that friendships and partnerships, such as those between J.R.R. Tolkein and C.S. Lewis (Glyer 2007), Joseph Conrad and Ford Madox Ford (Farrell 2001) and Charles Lamb, Samuel Taylor Coleridge and William Wordsworth (James 2008), play in the compositional process. Taken together these relationships reveal something about the distribution of agency across the scene of writing and about what these relationships produce in terms of shared or similar literary forms. Yet they reveal comparatively little about the spatial happening of these relationships. These writers know each other, they may live near one another, they may write to one another or attend the same literary salons and their work may exhibit shared features, and it is these factors that are often taken as indicators of collaborative or social making. They occur though not as practices, but as objective givens and so we come to know little about what actually happens – the lived sociality – within these moments or spaces of encounter. What is more, too frequently sociality is understood as something quite public and professional. It is the involvement of editors or other professional writers, and we fail to appreciate how literary practice is more intuitively and spontaneously social. Writers are children, partners, parents, colleagues and friends, to name but a few of their social ties and these bonds, while often unacknowledged, can often be a formative part of the creative process. Thus, to develop a fuller understanding of writing's sociality, we need to attend not only to the presence of others, but to the practice, both spatial and social, professional and ordinary, of these relationships: to how they happen, unfurl and go along within the world.

Other ways of moving closer towards the practices before the text present themselves within textual geography, where particular attention has been given to the study of eighteenth-century geography books, grammars and maps. Drawing on a range of geographical manuals, Miles Ogborn (2004) examines how such texts depended upon particular inscription practices, where choice of hand, style and decoration were central to the shaping of geographical knowledge. Although the emphasis is upon what such practices achieved within the public sphere, there is an implicit recognition of how penmanship emerged from a particular matrix of social anxieties and expectations. This matrix of social relations did not, however, always precede writing, for as Ogborn (2007) demonstrates in a study of the textuality of the East India Company, collective decision-making, civility and authority were also made through writing and inscription practices. This lengthens our view of the scene of writing, for it suggests that literary practice does not stop with the final flourish of pen on paper and nor is it reducible to the social relationships that produce this flourish. Instead, we need to understand literary practice and the writings that comprise it as something that continually goes along, both effectively and affectively, within a social world. Writing is not only made but it *makes*. It makes social relations, authority and knowledge, but we must not overlook how writing makes writing. This may suggest intertextual practices or the significance of draft materials, but it is more than this; it is the way in which writing becomes a way through. We use writing as a conceptual tool; it is a means of thinking, creating, and developing ideas. Thus, each act of writing can become a *writing-through:* a way of elucidating thought and understanding within a larger and longer process of literary making.

The go-along nature of writing is captured by various geographers, who recognise the role of going, seeing, recording and revising in the development of geographical knowledge. The notebook, the field journal (Lorimer and Spedding 2005) and the letter (Ogborn 2007; Withers 2004) are all moments within a larger scene of practice. They are ways of gathering and gleaning information, recording experience in-situ and they operate as aide memoires, travelling from the field to the laboratory or study where they may inform and shape subsequent textual practices. Two important issues arise here. Firstly, this body of work illuminates the iterative and reiterative nature of creation; it is a succession of tracks back and forth, between field site and study, between one individual and another and between one textual making and another. If we transfer these ideas to literary practice, we must recognise the scene of writing not as something linear and straightforward, but as something composed of turns and returns, convergences and divergences, iterations and reiterations. Secondly, textual geographers reveal how ideas travel and transform as they go along. Whether it is from field site to laboratory or from one side of the world to another, the meaning and materiality of the written word is affected by the social spaces it emerges from and moves in and through. Being *en route* or being in motion therefore, is recognised as crucial to the making of geographical knowledge. The significance of this state of being to literary practice is somewhat more obscure. While we recognise the role that place plays in inspiring the creative process (Brace and Johns-Putra 2010) we have yet to fully understand the roving nature of this creativity, or what Pierre Bourdieu (1977) might term, the very movement of its accomplishment.

One reason, perhaps, for the paucity of interest in the *en route* nature of literary prose lies in its inherent fictionality and the associated idea that what literature knows is always non-referential to people or place: it is subjective, unspecific and unverifiable, and so its meaning does not rest on its verisimilitude to the world. As such, there is no clear relationship between a writer's travels and the wor(l)ds they write. This is, however, reminiscent of a 'stamp-collecting' approach (Thrift 1978) to the relationship between literature and geography, which works only with the finished text and reduces geography to little more than points on a map. Yet literary writing and geography are bound in more inextricable ways, with John Wylie (2010), in a very self-reflective piece, exploring how landscape and self mingle and merge in a constant process of poetic making. Although this presses us to consider whether we can grasp the *en route* nature of literary practices that are not our own (a point discussed below), it is a work that recognises that unlike the scientist, the writer has no easily definable site of study and no prescribed ways of going forth or collecting 'data'. Instead, they are swept up in what Tim Ingold (2011) terms, the generative currents of the world and it is by moving with these currents and pressing on with their lives that making emerges. This suggests that it is not always specific landscapes or places that inspire or motivate literary practice, but rather, 'the episodes and occurrences of everyday life' (Saunders, 2010: 444). Literary practice continuously unfurls and happens through and during our lived experiences and understanding this duration often depends upon knowing a writer as 'more-than' a writer. As Brace and Johns-Putra (2010) observe, it involves understanding the beings and doings of authorship: what it means to be a writer beyond the physical act of putting pen to paper or finger to keyboard. Put another way, it pivots upon understanding a writer's geography as something that exceeds the spatial references and inferences of their texts.

Scenic methods: practice, biography and history

In drawing attention to the being and doings of authorship, Brace and Johns-Putra (2010) recognise the lived presence behind or responsible for the written wor(l)d. It is a presence they explore through 'talking-with' writers about the making of their work-in-process. Although talking-with may appear to come after rather than within the moment of accomplishment, it is part of the being and doing of contemporary authorship. Problems inevitably arise though, when we move away from contemporary writers to explore past literary practices. Here we have few opportunities to 'talk with' the writer, and so we find ourselves in the position of having to explore literary practice in the material traces authors leave behind them. These traces are frequently textual and relying on them to reveal authorial doings may seem a little paradoxical within the currents of practice and non-representational thinking. It is not as strange as it seems though, for as historical and cultural geographers point out, texts are not staid holders of meaning, they are lively and inventive assemblages (Griffin and Evans 2008; Ogborn 2009). Most obviously perhaps, they are lively at the moment of reception as readers reanimate and remake textual meaning (Hones 2014). Yet this reception can also be more self-referential as authors return to and revise their textual workings. Texts teem with life in other ways too; in their language (Evans 2008), in the loop, wobble and oscillation of their handwriting (Ingold 2011), and in their style (Travis 2008) and, following work in material culture studies that attend not to the material object *per se*, but to the properties that comprise it, it is possible to comprehend the text as an assemblage of movements, relationships and practices

(Ingold 2011; Patchett 2008). Such comprehension, as Merle Patchett's study of taxidermy and Caitlin DeSilvey's (2007) work with the redundant and decaying objects of a Montana homestead demonstrate, requires a degree of expressive and inventive researching, what Hayden Lorimer (2006) might term, 'make-do' methodologies. This can involve immersive ethnographies, the retracing and reanimating of lived cultures and the interpolation of material traces, which when read and explored together move us closer to the liveliness of the past. It is a version of this latter approach that is developed here. Drawing on the work of Nedra Reynolds (2004), who argues that writing is a spatial practice that overlaps and fuses with other spatial practices, such as inhabiting, dwelling, moving and mapping, this book argues that practices invariably nest within and disclose one another. Textual traces register our being-within-the-world, they capture our whereabouts, our travels, and our general doings, and extrapolating these nestings enables us to explore both the spatial and social expansiveness of the scene of writing, for material traces are pregnant with, and exist as sites of lived encounters and social interactions (Saunders 2008; 2010).

Thus, this book works from the premise that we can access past literary practices through the textual traces of an author's life; their letters, journals, memoirs, writings and diaries. Such an approach inevitably draws us towards questions of biography, a subject many disciplines view with suspicion, not least because it tends to fix and isolate a subject in place, essentialising it as a coherent entity untouched by other lives (Livingstone 1992; Stanley 1995). If we are to understand, more fully, the spatial and social practices at play within the scene of writing, we need to reanimate the author, but this need not be a biographical enterprise in the traditional sense. Not only because biographical studies are increasingly recognising complexity over coherence and celebrating the interdependencies and kaleidoscopic nature of multiple lives, but because geography offers us a way of examining what Livingstone (2003) terms, the spaces of a life. This is how different spaces foster different selves and provide opportunities for the plotting of different life stories (Daniels and Nash 2004). Thus, biography is not a teleological progression as such, but rather, an amalgam of different selves that both co-exist and come to the fore at different times. Biography need also not be a study of that which is thought big. Lorimer (2003) alerts us to the importance of small stories, to those everyday and mundane happenings that frequently fall out of the purview of biography, but which are no less constitutive of how we know and encounter the world. Innes Keighren (2007), for instance, urges us not to ignore the daily, mundane detail of life, for while this may not explain particular engagements with the world it can facilitate them. This is important in another way, for it moves us away from thinking of biography as explanation and, instead, makes us think of biography as constitutive; biography is not a series of ready-made accomplishments, but is always a lived process of making.

An enlivened biographical lens, therefore, has the scope to open up the scene of writing in a different way. It enables us to see the compositional process not as a linear undertaking, in which others exist purely in the shadows and engage only at set moments, in certain ways and with certain intentions, but as a more discontinuous, spontaneous and everyday practice. It is more akin, perhaps, to Ingold's (2011) idea of the meshwork, which conceives of the world as a tangle of life lines – of people, places and things – that mingle, connect and leave some impress of themselves on one another before moving on. The meshwork is always more than a touching of discrete lives; it is an intertwining of lives that has the capacity to affect the trajectory of a life's line and, by implication, the nature of the compositional process. Considering the scene of writing as a meshwork has other implications too. It uncouples author and text and allows us to see Anthony Trollope, John Galsworthy and Arnold Bennett as more-than authors, or to see their literary practices as something more than a responsibility for the written word. As a meshwork, the scene of writing is not those doings that lead directly to the text, but rather, the mundane happenings of life: the domestic toils and travails, the programmes of work begun but never finished, and the public roles and bread-and-butter employment that make writing feasible and possible. The scene of writing, therefore, is one of conscious and unconscious activities, connections and reconnections, dead-ends, partial journeys, and tracks and back-tracks. Considering the scene of writing in this way offers another intervention into the relationship between literature and geography. It is an intervention that attends not to the geography in the text or the geography of the text *per se*, but to the geographies of everyday knowledge in which literary writing is implicated.

Talk of the author and of reanimating the author makes it pertinent to close with some reflection on the death of the author. Authorial studies have been, and to some extent still are, unfashionable. Genette (1980) observes that the world of the author is not a sufficiently rigorous source for illuminating textual meaning. This may well be true, but textual meaning is different to textual or literary practice. The latter is concerned with the interpretation of doing and not, as in the former, with the interpretation of what has been done. In the same way that Brosseau (1994) urged us, over 20 years ago, to listen to the text, so it is time that we listened to, or perhaps more aptly, engaged with, the author as a maker and the process of their making as something that can be, partially at least, recovered. At a time of growing interest in geographies of craft and making, and when geographers have begun to problematize their own literary practices (DeLyser and Hawkins 2013), it could not be timelier or more urgent that we bring ideas of creative doing and skilled practice to literary geography and literary studies more broadly.

References

Attridge, D. 2004. *The Singularity of Literature*. London: Routledge.

Blair, S. 2004. Local modernity, global modernism: Bloomsbury and the places of the literary. *English Literary History*, 71 (3), pp.813–838.

Bourdieu, P. 1977. *Outline of a Theory of Practice*. Cambridge: Cambridge University Press.

Brace, C. and Johns-Putra, A. 2010. Recovering inspiration in the spaces of creative writing. *Transactions of the Institute of British Geographers*, 35 (3), pp.399–415.

Brosseau, M. 1994. Geography's literature. *Progress in Human Geography,* 18 (3), pp.333–353.

Bushell, S. 2009. *Text as Process: Creative Composition in Wordsworth, Tennyson, and Dickinson*. London: University of Virginia Press.

Chambers Dictionary, The 1998. New ed. Edinburgh: Chambers Harrap.

Daniels, S. and Nash, C. 2004. Lifepaths: geography and biography. *Journal of Historical Geography*, 30 (3), pp.449–458.

Dawson, P. 2005. *Creative Writing and the New Humanities*. London: Routledge.

DeLyser, D. and Hawkins, H. 2013. Introduction: writing creatively – process, practice, and product. *Cultural Geographies*, 21 (1), pp.131–134.

DeSilvey, C. 2007. Salvage memory: constellating material histories on a hardscrabble homestead. *Cultural Geographies*, 14 (3), pp.401–424.

Doyle, C. 1998. The writer tells: the creative process in the writing of literary fiction. *Creativity Research Journal*, 11 (1), pp.29–37.

Evans, A.B. 2008. Enlivening the archive: glimpsing embodied consumption practices in probate inventories of household possessions. *Historical Geography*, 36, pp.40–72.

Farrell, M. 2001. *Collaborative Circles: Friendship Dynamics and Creative Work.* Chicago: Chicago University Press.

Freeborn, R. 1960. *Turgenev: The Novelist's Novelist.* Westport: Greenwood Press.

Genette, G. 1980. *Narrative Discourse: An Essay in Method.* New York: Cornell University Press. Trans. J. Lewin.

Glyer, D. 2007. *The Company They Keep: C.S. Lewis and J.R.R. Tolkien as Writers in Community*. Kent: Kent State University Press.

Griffin, C.J. and Evans, A.B. (2008), On historical geographies of embodied practice and performance, *Historical Geography* 36, pp.5–16.

Heidegger, M. 1978. *Being and Time*. Oxford: Blackwell.

Hones, S. 2014. *Literary Geographies*. London: Palgrave Macmillan.

Hones, S. 2008. Text as it happens: literary geography. *Geography Compass,* 2 (5), pp.1301–1317.

Ingold, T. 2011. *Being Alive: Essays on Movement, Knowledge and Description.* London: Routledge.

James, F. 2008. *Charles Lamb, Coleridge and Wordsworth: Reading Friendship in the 1790s*. Basingstoke: Palgrave Macmillan.

Jones, R.E. [1941] 2004. *The Dramatic Imagination.* London: Routledge.

Keighren. I. 2007. Breakfasting with William Morris Davis: everyday episodes in the history of geography. In E. Gagen, H. Lorimer and A. Vasudevan, eds. *Practising the Archive: Reflections on Method and Practice in Historical Geography.* London: Royal Geography Society/Historical Geography Research Group, pp.47–56.

Kneale, J. 1999. The virtual realities of technology and fiction: reading William Gibson's cyberspace. In M. Crang, P. Crang and J. May, eds. *Virtual Geographies: Bodies, Spaces and Relations.* London: Routledge, pp.205–221.

LeFevre, K.B. 1987. *Invention as a Social Act.* Carbondale: Southern Illinois University Press.

Lehan, R. 2005. *Realism and Naturalism: The Novel in an Age of Transition.* Madison: The University of Wisconsin Press.

Livingstone. D. 2003. *Putting Science in its Place: Geographies of Scientific Knowledge.* London: Chicago University Press.

Livingstone, D. 1992. *The Geographical Tradition: Episodes in a Contested Enterprise.* Oxford: Wiley-Blackwell.

Lorimer, H. Spedding, N. 2005. Locating field science: a geographical family excursion to Glen Roy, Scotland. *The British Journal for the History of Science*, 38 (1), pp.13–33.

Lorimer, H. 2006. Herding memories of humans and animals. *Environment and Planning D: Society and Space*, 24 (4), pp.497–518.

Lorimer, H. 2003. Telling small stories: spaces of knowledge and the practice of geography, *Transactions of the Institute of British Geographers*, 28 (2), pp.197–217.

Love, H. 2002. *Attributing Authorship: An Introduction.* Cambridge: Cambridge University Press.

Lubbock, P. 1921. *The Craft of Fiction*, London: Jonathan Cape.

Lunsford, A. Ede, L. 2012. *Writing Together: Collaboration in Theory and Practice*. New York: Bedford St Martin's.

McGann, J. 1991. *The Textual Condition.* Princeton: Princeton University Press.

Merleau-Ponty, M. 2013. *Phenomenology of Perception.* London: Routledge.

Noxolo, P. and Preziuso, M. 2013. Postcolonial imaginations: approaching a 'fictionable' world through the novels of Maryse Condé and William Harris. *Annals of the Association of American Geographers*, 103 (1), pp.163–179.

Ogborn, M. 2009. Francis Williams's bad language: historical geography in a world of practice, *Historical Geography*, 37, pp.5–21.

Ogborn, M. 2007. Indian Ink: Script and Print in the Making of the English East India Company. London: University of Chicago Press.

Ogborn, M. 2004. Geographia's pen: writing, geography and the arts of commerce, 1660–1760. *Journal of Historical Geography*, 30 (2), pp.294–315.

Parker, H. 1983. Non-sense, skewed meanings and intentionless meanings. *Critical Inquiry*, 9 (4), pp.767–774.

Patchett, M. 2008. Tracking tigers: recovering the embodied practices of taxidermy. *Historical Geography*, 36, pp.17–39.

Pavis, P. 2012. *Contemporary Mis En Scène: Staging Theatre Today*. Abingdon: Routledge. Trans. J. Anderson.

Reynolds, N. 2004. *Geographies of Writing: Inhabiting Places and Negotiating Difference*. Carbondale: Southern Illinois University Press.

Reynolds, N. 1993. Ethos as location: new sites for understanding discursive authority. *Rhetoric Review*, 11 (2), pp.325–338.

Saunders, A. 2010. Literary geography: reforging the connections. *Progress in Human Geography*, 34, pp.436–452.

Saunders, A. 2008. The practice of the text: arts of conversation, arts of writing. *Historical Geography*, 36, pp.94–111.

Shiach, M. 2005. Modernism, the city and the 'domestic interior'. *Home Cultures*, 2 (3), pp.251–267.

Shotter, J. 1993. Conversational Realities: Constructing Life through Language, London: Sage.

Shusterman, R. 2001. Art as Dramatisation. *Journal of Aesthetics and Art Criticism*, 59 (4), pp.363–72.

Stanley, L. 1995. Romantic friendship? Some issues in researching lesbian history and biography, *Women's History Review*, 1 (2), pp.193–216.

Taplin, O. 2002. *Greek Tragedy in Action.* 2nd edn. London: Routledge.

Thacker, A. 2005. 'Mad after foreign notions': Ezra Pound, imagism and the geography of the Orient. In P. Brooker and A. Thacker, eds. *Geographies of Modernism: Literatures, Cultures, Spaces*. London: Routledge, pp.31–42.

Thrift, N. 1978. Letters to the editor. Landscape and literature, *Environment and Planning A,* 10, pp.347–349.

Travis, C. 2008. Beyond the Cartesian pale: travels with Samuel Beckett, 1928–1946. *Historical Geography*, 36, pp.73–93.
Withers, C. 2004. Writing in geography's history: Caledonia, networks of correspondence and geographical knowledge in the late enlightenment. *Scottish Geographical Journal*, 120 (1–2), pp.33–45.
Wittgenstein, L. 1973. *Philosophical Investigations*. Oxford; Wiley-Blackwell. Trans. G.E.M Anscombe.
Wylie, J. 2010. Writing through landscape. In C. Brace and A. Johns-Putra, eds. *Process: Landscape and Text*. Amsterdam: Rodopi, pp.45–64.
Yapp, E.X.Y. 2011. Readers-in-conversation: a politics of reading in literary geographies. *Social and Cultural Geography*, 12 (7), pp.793–807.

Part I

The place of writing

1 Interpretations on an interior*

> 'Place is not mere background atmospherics, but provides for the very possibility of intellectual innovation. Intellectual inquiry is not the view from nowhere, but the view from somewhere.'
>
> (Barnes 2004: 568)

Histories of geography have long been interested in the relationship between the place of the discipline's happening and the texts that get written within and about these places. The contention among geography's historians is that the *where* of geography matters to the *what* of geography (Barnes 2004; Livingstone 2003; Lorimer and Spedding 2005). It is a body of work that is inspired by Bruno Latour's (1987) research on science in action. Latour's premise is that science comes to us as a smooth, coherent, rational entity. It is, to borrow another of Latour's terms, a black box: a reasoned, given whole that masks its own practice, its own social genealogy. As a result, we are accustomed to think of scientific knowledge as universal, replicable and categorical. If, however, we explore the practices that go into the making of this black box, including the spatialities of these practices, the whole nature of scientific knowledge takes on a different hue. We begin to see it as something made by, and contingent upon, the specificities of its place: institutional norms, the patterns of power, networking and alliance-building, and the exchange and construction of information, reveal science's making to be chaotic and ragged, and its knowledge to be far from certain or rational.

The purpose of this chapter is to pose a very similar set of questions in relation to the practice of imaginative forms of writing. As argued elsewhere, the doing of science and the doing of imaginative literature have different relationships to place (Saunders 2010). Where the former regards 'being-in-place' and 'seeing-for-one's-self' as bound up with the production of truth and veracity, the latter positions the relationship between the real and the imagined as one of authenticity rather than verisimilitude. In consequence, to ponder the place of literary practice is not to ask, as early work in literary geography did (Darby 1948; Gilbert 1960), how closely the real aligns with the imagined, but rather, to consider how the places of literary practice function within the creative process. Thus, while this

*An earlier version of this chapter appeared in Literary Geographies, 1 (2), pp. 174–94.

book argues that writing creatively is more than what goes on in those places where pen is put to paper, this chapter (and Part I as a whole) contends that we must not wholly overlook these places, for where pen is put to paper is rarely an isolated garret or ivory tower; rather it happens within lithe social worlds. Spending time in these places and attending to their materiality – to what winds up in these places, to who gathers within them, and to how they are laid out and designed – discloses something of the way in which place, and the things that happen within place, matter to the happening of imaginative writing.

It is to these questions that this chapter turns, taking as its starting point the material world that was inhabited by the English writer, Arnold Bennett (1867–1931). Bennett had a varied literary career, and while he may not be widely known today, he was one of the most popular and successful writers of his day (Drabble 1975). In 1914 Bennett published *The Author's Craft*, a guide on how to write fiction. A central element within this work was the importance that Bennett attached to geography as a prelude to writing. Ostensibly, he meant a cognisance of place as setting, but of particular interest here is the persistent significance that Bennett attached to the organisation of his own place of writing: to the assembly, fashioning and arrangement of his domestic interiors and the way in which these mattered to his literary practice. Before exploring these interiors and what they might have meant to Bennett's literary making, let us turn to examine the broader relationship between place and literary creation.

Literary interiors

This chapter is not the first to turn its attention to the place of imaginative writing. In *The Senses of an Interior*, Diana Fuss (2004) probes the relationship between interior spaces, both psychological and architectural, and creative lives. The premise of Fuss's text is that the where of literary labour is of central importance to the nature of its happening and its meaning. It discusses the work of four writers, Emily Dickinson, Sigmund Freud, Helen Keller and Marcel Proust, whose literary lives span the 1850s to the 1960s. This period is significant, for whether they were engaged in poetry, prose or psychoanalysis these writers were performing their work at a time when the interior, as a space apart from the exterior public world and as a place of mental dwelling, was under construction (Benjamin [1930] 1999; Rice 2007). Such that, prose of the interior was simultaneously prose of interiority; rendering the domestic interior in painstaking detail was one means by which writers sought to explore and reflect on their mental interiors. In consequence, Fuss's work 'opens a window onto . . . author[s] and text[s], reminding us that what we may at first perceive to be the timeless and universal truth of writing cannot be so neatly extricated from the complex particularities of spatial and material origins' (2004: 2).

Fuss's interest in the interior is part of a resurgent interest in the world that comes before the text. It is a world that, for much of the twentieth century, remained outside the purview of literary scholars. The influence of the Romantic Movement, with its emphasis on creativity as the work of a lone consciousness – and therefore, elusive, ephemeral, and unknowable – remained strong. Alongside this, New Criticism with its rejection of authorial intentions and, more recently, post-structuralism's pronouncement of the death of the author directed attention away from the world of the writer and towards the world of the text (Barthes 1977; Wimsatt and Beardsley 1946). Fuss's work, however, is part of a broader attempt to recover the world of the writer as something more than a footnote to textual meaning. Literary geographers bring a particular dimension to this recovery attempt, for as Hones' (2014) work demonstrates, they are interested not just in place as a container of things, but in its making and moving. Drawing inspiration from Doreen Massey's (2005) conceptualisation of space as a product of interrelations, a dimension of coexistence and a process of becoming, Hones uses this concept to argue for the novel as both an outcome and an on-going process of intricate spatial relationships. Thus, what literary geographers bring to studies of the literary interior is a cognisance of the relationality of interior space. It is not enough to know what occurs therein; we also need to know how these spaces are themselves produced and through what kinds of social and spatial relationships. The interior is not just a stage; it is interwoven into the being and the doing of the writer. Writing, as already noted is a process of *longue durée*; and, as Brace and Johns-Putra observe, 'being a writer goes on beyond the act of writing and also occupies spaces other than those in which the writing goes on' (2010: 411). Hence, while this chapter is interested in the place where pen and paper meet it recognises that what happens in the study or writing room is not always easily uncoupled from what happens in the other social spaces a writer inhabits in their day-to-day life. To draw out these ideas it is time to turn to the life and work of Arnold Bennett, and examine his literary interiors through a more explicitly geographical lens.

The writers that guide Fuss's examination of the interior were living and writing at a time when the nature of the interior was undergoing mental and material transformation. Arnold Bennett's writing career fell firmly within the middle of the period Fuss considers, beginning with the publication of *A Man from the North* in 1898. However, unlike the writers Fuss considers, Bennett's work is rarely noted for its psychological depth or stylistic innovation. Indeed, he is a writer often placed 'outside modernism'; one who did not experiment with a turn inwards towards interior consciousness and an identity-based political aesthetic (Ardis 2002; Paxton 2000)[1]. Nor was Bennett noted for the quirkiness of his interiors in the way that Freud or Proust were. Despite being one of the foremost writers of his day, few photographs exist that document the various homes Bennett inhabited. That said, Bennett's interiors were subject to another form of critique. In 1924 Virginia Woolf's now famous essay, *Mr. Bennett and Mrs. Brown* ([1924] 1959), appeared under the imprint of the Hogarth Press. This essay, the development of a number of earlier pieces, took Bennett to task for the nature of his prose, which was, Woolf argued, comparable to well-built houses in which nobody lived. Woolf carried this domestic analogy further, deriding what she saw as Bennett's detailism: his tendency to describe character through a detailed inventory of houses and homes, which in its focus on material facts missed life itself.

The quarrel between Bennett and Woolf over aesthetics has been explored extensively elsewhere (Castle 2015; Squillace 1997); yet it is worth reminding ourselves of its lineaments not least for the way it was impelled by different senses (and values) of the interior. Woolf's main criticism of Bennett focused on the psychological weakness of his prose; he rarely got beneath the skin of his characters. Thus, we know much about the detail of their lives, about what they looked like and where they lived, but when we try to move beyond the solidity of their lives, to enter their homes or their minds, the characters crumble and fade: there is nothing beyond the material world to get hold of. Bennett, in turn, was critical of Woolf's retrenchment of the lived in favour of what he termed 'fancy' and 'padding' so that characters never seemed quite present (1929: 5). Where spatial interiority is a cipher for psychological interiority in Woolf's thinking, for Bennett the spatial interior is a way of narrating the self and disclosing its multi-faceted nature.

It is unsurprising, then, that houses are afforded such centrality within Bennett's fiction; although Woolf's critique was inspired by *Hilda Lessways* (1911), it is *Clayhanger* (1910) that develops through a series of houses both material and imagined. The novel follows Edwin Clayhanger's interest in the building of his family's new home, 'to Edwin it was not a house, it was a work of art, it was an epic poem, it was an emanation of the soul' (Bennett [1910] 2000: 169) and it was an emanation that altered his very perception of architecture:

> [He] had always looked on a house as a front-wall diversified by doors and windows, with rooms behind it. But when Mr. Orgreave produced his first notions of the new house Edwin was surprised to find that he had not even sketched the front. He had said, 'We shall be able to see what the elevation looks like when we've decided the plan a bit.' And Edwin saw in a flash that the front of the house was merely the expression of the inside of it.
>
> (Bennett [1910] 2000: 170)

Houses, for Bennett, were not a substitution for character; they were part of one's character and life. Inspired, to some extent, by naturalism, which regarded the material world as an inextricable influence on character and action, Bennett saw where one lived, how one lived and how one arranged one's material possessions to be expressive of personality, identity and self-development (Lehan 2005). This aesthetic idea was not confined to Bennett's imagined world; it was, as we shall see, a formative influence on his domestic world.

Assembling the interior: Bennett's home-making practices

In 1903 Bennett moved from London to Paris, believing the latter city to be a more conducive environment for creativity. Soon after his removal to Paris, Bennett wrote to his friend and fellow novelist, H.G. Wells, observing that 'I have got a charming little flat here, & furnished it myself' (8 October 1903, in Bennett 1968: 182). Throughout much of 1904 he records regular trips to 'my "Empire" shop', buying on one occasion 'two occasional tables, a candlestick, and a flower-glass, all strictly Empire. I have now done buying furniture. I only want bibelots and things' (2 June 1904, in Bennett 1932: 178). Empire furniture took its name from the Napoleonic Empire of 1804–1814. In style it was imposing and opulent, heavy and dark. It was often made from mahogany or ebony and richly patterned with symbols and motifs. Undoubtedly this was an interior rich in signification, but what is also apparent is Bennett's care for his interior. It was a care that was ostensibly material, for Bennett's interiors arose not through random happenstance but through painstaking assembly as objects were identified and arranged in-line with an overriding aesthetic. Yet, the very process of assembly discloses a pre-condition of dispersal and disassembly, and in navigating between the two, the cares of Bennett's material world fold themselves together with those of his mental world. It is at this moment of enfolding as one world intrudes upon and potentially upsets the other, that we catch up with Arnold Bennett and consider the implications this doubling of the interior – as material and mental space – has upon his writing practice.

It is to the Villa des Néfliers, the house Bennett rented in Avon-Fontainebleau soon after his marriage to Marguerite Soulié in 1907, that we turn first. Bennett had begun searching for a new home in the autumn of that year, visiting Avon-Fontainebleau one wet, November day. Despite the weather he took 'distinct pleasure in examining [. . .] [the] houses' and quickly 'fell in love with the one I liked, and at once, in my mind, arranged it as it ought to be' (26 November 1907, in Bennett 1932: 269). Bennett and Marguerite took up residence at Les Néfliers in April 1908, but it took Bennett several weeks to organise the house to his taste: 'I haven't yet arranged my days here. I am doing no reading, no fine writing, no disciplinary thought of any kind. It is true that I still spend about two hours a day in working at the arrangement of the house' (2 May 1908, in Bennett 1932: 288). Throughout May the interior arrangement of Les Néfliers came to disrupt Bennett's working routine. On 17 May 1908 Bennett and Marguerite returned home to find the 'house overrun with ants'. While this invasion caused some disquiet, it was the subsequent arrival of new carpets which, while 're-arousing our pride in our toy house', required Bennett to forego his 'afternoon sleep in order finally to arrange the second spare room' (17 May 1908, in Bennett 1932: 290). The full import of such seemingly minor events becomes evident as the month wears on:

> To-day I seemed to get a little nearer the state of mind and the mode of life that I have aimed at [. . .]. I have finally got my brain far better under control [. . .] [but I am] haunted by dissatisfaction at the discrepancy between reason and conduct! No reason why conduct should not conform to ideas of reason, except inefficient control of the brain. This I am always preaching, and with a success of popular interest too, I cannot perfectly practise. It is the clumsiness of my living that disgusts me. Half an hour in the morning in complete concentration on the living-through of the day, and I should work wonders! But this all-important concentration is continually interrupted – interruptions

which weaken it; sometimes deliberately abandoned for concentration on matters of admittedly interior importance.

(23 May 1908, in Bennett 1932: 291–292)

Les Néfliers is an interior in the making and one whose making unmakes and disarranges Bennett's mental interior. The process of making the interior is very much one of assembly that involves removal companies, carpet fitters, self-reflection and personal action, not to mention the local wildlife and the temporalities of the building itself.

Exploring Les Néfliers through the lens of assemblage thinking, as a space in composition and therefore in a perpetual state of experimental flux (Anderson et al. 2012), gives insight into the transformative and performative nature of the interior. Assemblage thinking is increasingly used within the social sciences to explore the world as process rather than product, as something in the making rather than as something made. Thus, assemblage approaches reject conceptions of the world as a set of pre-given properties and binary entities (such as inside/outside), and instead, turn their attention to the often messy ways in which people, things and processes gather and assemble in the moment (Anderson and McFarlane 2011). What this means, in the context of Les Néfliers, is that the house was not a static container into which the Bennetts simply moved. They may well have imagined it to be an ordered idyll, a perfect place for artistic and married life; but through the process of moving and making it becomes in ways not necessarily anticipated. Thus, it was not a blank slate awaiting inscription, but a restive, resistive agent. Thomas Gieryn (2002) has suggested that buildings in the process of assembly are symbolically weak structures. The process of making generates so many different possibilities, discontinuities and opportunities that buildings have little power to order or express social (or mental) life. The same could be said of a building's interior; as yet it symbolises nothing but itself, although it proffers many opportunities for development. Thus, the on-going (and for Bennett the frustratingly unfinished) nature of Les Néfliers – the lack of order, of things not out of place as such, but as yet having no place and the process of finding places – is, to borrow from James Ash's (2009: 2109) work on screen geographies, 'productive of its own . . . way of creating attention and bringing different worlds into being'. These are not the worlds Bennett necessarily wants, but in making them possible the process of assembly proffers previously un-thought possibilities.

Let us turn at this point from Bennett's unfinished domestic interior to his mental interior, for the two are, as Fuss (2004) and Charles Rice (2007) demonstrate, intimately enfolded in one another. The disruptiveness of Les Néfliers could be taken as little more than procrastination on Bennett's part, although the wider point is that place and the (dis)organisation of place matters to the very happening of practice. Equally important though, is what Bennett was actually practising at this time. Since the middle of 1907 he had been working on the novel that would become *The Old Wives' Tale* ([1908] 2007), and which would be published in the autumn of 1908. Bennett always liked to break up his longer compositions with briefer works, such as articles and short stories, but during the writing of *The Old*

Wives Tale he was, as Margaret Drabble (1975) observes, particularly productive. He wrote two short novels, *Helen of the High Hand* (although this was not published until 1910) and *Buried Alive* (1908), multiple articles and short stories, he contemplated a book of poetry, finished writing a play with the novelist Eden Phillpotts, saw his play *Cupid and Commonsense* appear on the London stage and wrote a couple of books of popular philosophy. In a sense, what transpires in this period of domestic disassembly and assembly is an intensified mental flitting from one project to another; an inability to settle to any one task for more than a moment, with the result that many different imagined worlds are begun and worked upon, with some coming to fruition and others being put to one side. It was only at the end of May, when Bennett believed he had organised his house as he wished, that he returned to *The Old Wives' Tale*, completing it in August of that year. What this suggests is that the assembly of one interior is productive of the disassembly of another; putting his home in order dispersed Bennett's concentration, inhibiting sustained work on his longer-term projects but facilitating those shorter texts that rarely required more than a few days of work. Thus, interiors in-assembly may disassemble others, but in so doing they can give birth to alternative assemblages and ways of going-forth.

Fashioning the interior: the play of influence

Bennett's fastidiousness in respect to interior design was of relatively long standing. At the turn of the century, while still living in Britain, Bennett had taken out a lease on Trinity Farm. The farm, located in Hockliffe, Bedfordshire, was Bennett's measure of his own literary success. Until 1900 he had been editor of *Woman* magazine as well as writing a number of short stories, essays and his novel, *The Man from the North* (1898). It was his growing literary success that encouraged him to give up the editorship in 1900 in order to turn his hand to writing full-time and freelance. A house in the country seemed to offer him all the accoutrements of a successful writer at this time. John Galsworthy, for instance, was to rent a country house in Manaton, Devon; H.G. Wells had a home in Sandgate, Kent; Eden Phillpotts lived in Torquay; and Thomas Hardy lived just outside Dorchester, Devon: country living was in vogue. Bennett knew all of these writers, either personally or by reputation, and he emulated their lifestyle and their determination to remove from London in order to write (Drabble 1975). The spaciousness of Trinity Farm also enabled Bennett to provide a home for his ailing parents, and a refuge for his recently bereaved sister. During his editorship of *Woman*, Bennett had become familiar with the currents of fashionable design and had developed very particular artistic ideas. What is more, he was unaccustomed to compromise, becoming very fastidious about certain aspects of the farm's interior decoration:

> With regard to the papering, it is understood that I should be at liberty to spend 2/6 per piece for the living-rooms and 1/6d for everything else. It is absolutely essential that I should be at liberty to choose my own patterns from the firm of Essex & Co., in Victoria Street [. . .] I am extremely particular

> about the wall-papers that I have to live with, & only this firm sells the artistic patterns which I require.
>
> (24 August 1900, in Bennett 1968: 136)

We will return to this artistic wallpaper in a moment, but what is emerging from this letter and from Marguerite Bennett's (1925) recollections of Bennett's obsession with interior design, is a picture of a writer quite intransigent over his interior and used to getting his own way. Another instance of this stubbornness arose when Bennett cohabited with his sister Tertia and brother Septimus in Fulham. The three shared a house, paid for by Bennett, from 1897 to 1900, and in a letter Bennett wrote to his friend John Rickard a sense of his control is clearly evident: the house's glory is its 'two studies . . . one for poetry, the other for prose! Try to grasp that. They wanted to seize one of the studies as a place convenient to put a sewing machine in! God! I said "No" to that' (8 December 1897, in Bennett 1968: 95). Although Bennett's correspondence suggests his single-minded control over the unfolding of these spaces, it is also clear that numerous others were implicated, often in very material ways, in the fashioning of these interiors.

Houses have long been seen as sites of what we might, after Nicolas Bourriaud (2002), call relational aesthetics: spaces in which art is made and encountered in highly social ways. Within London, houses frequently played host to literary salons where writers could share ideas, network with publishers and benefit from advice and even financial support (Vadillo 2007). Outside the metropole residential hubs developed wherein writers could come together as friends rather than professionals. One such was H.G. Wells' house in Sandgate. This was part of a network of houses through which a loose association, subsequently termed the Romney Marsh or Rye Circle (Farrell 2001), was sustained. Alongside H.G. Wells, this group included Henry James, Joseph Conrad, Edith Nesbitt and Stephen Crane. These writers visited one another, hosted literary notables and together came to exert a significant influence on the development of a modernist aesthetic. At Trinity Farm, however, there was another kind of relational aesthetic at play, what Jenny Sjöholm (2013) might call 'invisible socialities'. These are the wider frames of reference, expectation and influence that seep into and shape not only work practices but ideas of what a work place should be. It is a concept that draws on Pierre Bourdieu's (1993) notion of the spaces of possibles – the traditions, disciplinary problems and intellectual concepts that define and motivate a subject area. Inclusion (and success) within a subject field depends upon recognising and working with these prevailing ideas. Conventionally, 'invisible socialities' will manifest themselves in how a writer references past work or past masters, nods to particular traditions or demonstrates certain aptitudes and styles. It is also evident though in the way writers emulate, in very material ways, particular subject identities.

Let us return to Bennett's insistence on his particular choice of wallpaper from Essex and Co. We do not know what these artistic patterns were, but one of the chief and exclusive designers for Essex & Co. at this time was Charles Voysey.

Although perhaps best known for his architecture, Voysey had a broader design interest and his wallpaper designs were strongly influenced by both the art nouveau and the arts and crafts movement (Jackson 2007). Voysey's wallpaper came to be noted for its use of clear, bright colours, strong friezes and simple motifs (Latimer 1988) and has led some to suggest that his designs, like those of William Morris, were complicit in moving pattern design away from the historicism of the Victorian period towards flatter, less intrusive designs in which shadow and relief were absent and greater simplicity prevailed (Pevsner [1936] 1991; Vallance 1892). By 1896, as *The Studio* magazine observed, a ' "Voysey wallpaper" sounds almost as familiar as a "Morris chintz" or a "Liberty silk" ' (E.B.S. 1896: 209): it was shorthand for the height of interior fashion. Four years later, when Bennett was moving into Trinity Farm, Voysey was engaged in designing what was to become Spade House, for Bennett's friend H.G. Wells. This, along with Bennett's lifelong interest in design and his familiarity with the currents of interior fashion, given the centrality of these themes to the content of *Woman* magazine, make it highly probable that it was the work of this designer that he required. If this is the case, we begin to discern some of the invisible socialities present within Trinity Farm's interior. The material presence of these influences points to the way in which Bennett's interior is a negotiated space. It is one negotiated not through his most immediate or proximate relations, but rather through a relational aesthetic that is based upon an imagined version of the self and its inhabitation within a particular community of praxis. As Daniel Miller (2008) has suggested, the stuff of our homes, the things we collect therein are conduits to history and geography. They leapfrog the now and the immediate allowing us to conjoin ourselves to other times and places. Thus, choosing an interior scheme was not merely a question of securing what one liked, admired or found close-at-hand; it was, Ann Anderson (2013) points out, about gaining access to the imagined community of aestheticism: of being recognised as an individual (and an artist) of good taste. An interior, therefore, needed to signal belonging to a world that often lay far beyond its four walls or immediate neighbourhood; it needed to script one's credentials in highly material and tangible ways.

Not enough is known about the interior décor of Trinity Farm to push this relationship too far at present. What is known though, is that despite Bennett's insistence upon a particular aesthetic for the house, its making was not a straightforward process. Nearly two months after penning his letter to the letting agent for Trinity Hall, A. W. Merry, regarding his artistic wallpapers from Essex and Co., Bennett writes again to complain that the owner, Mr Adams, was proving difficult over the execution of his chosen interior:

> I chose the papers, & now it turns out that Mr. Adams wants more discount than Messers. Essex will give. Mr. Adams says that if I choose to pay the difference I can have the papers. This is of course ridiculous. The difference is some 6/- odd. The delay is getting serious [. . .] More than half the house is uninhabitable & the delay in every department of the repairs is gross.
>
> (12 October 1900, in Bennett 1968: 140)

This spat between Bennett and Adams crystallises two inter-related processes at play in the fashioning of the interior. Firstly, it points to how the interior is enmeshed within a power geometry (after Massey 1994)[2]. Most obviously it reveals Bennett's lack of currency as an author figure. Thirty years later, at the height of his fame, Bennett had interiors designed for him by Marion Dorn, one of the preeminent textile designers of the early twentieth century. In 1900, however, Bennett's authorial status did little to invite such favour; he was still a struggling writer, relatively unknown, particularly in the area in which he had chosen to reside, and as such unlikely to acquire the privileges that go along with celebrity. What is more, it was evident that having to pay the difference for his chosen wallpapers was going to be difficult. Thus, for those on the periphery of the artistic circles of the time, fashioning an artistic interior was a delicate undertaking. Interiors, as Rice (2007) argues, travelled, but they travelled most easily when assisted by reputation and income. Bennett had neither of these in abundance in 1900 and so the making of his interior was often one of compromise and collaboration, as factors other than the aesthetic came to fashion its development.

It is Bennett's friendship with H.G. Wells that reveals the second dimension of the interior's power geometry. Bennett's move to Trinity Farm has been seen as an attempt to emulate the modes of inhabitation common among his more successful literary contemporaries. Yet, as Drabble (1975) observes, the choice of Bedfordshire was an odd one, for in comparison to the other counties surrounding London there was little to recommend it socially or aesthetically. It is likely that the main determinant of Trinity Farm was cost comparative to the rest of London's rural hinterland, with the result, Wilfred Whitten records, that at Trinity Farm Bennett was between two worlds (Whitten in Bennett 1968: 139). He had not quite left the conventional middle-class home of his Staffordshire childhood, but nor had he fully entered the more progressive world of the literary avant-garde (139). The gentlemanly spat over the wallpaper is partly indicative of Bennett's hybridity within this cultural world, but so too is the choice of Bedfordshire: each was a compromise, but each was also part of a typology of belonging; another staging post within the development of an authorial identity. Trinity Farm and its interior became a space of organised striving in which competing ideas of authority were inscribed and played out: struggles arose, subject positions were moulded and attempts were made to close the distance between artistic worlds.

A sense of this striving is evident in Bennett's literary labours of the time and was most clearly expressed in a letter he wrote to the nature writer George Sturt several months after his arrival at Trinity Farm:

> I fancy I have sort of settled down to a literary life, pure & simple. My nearest approach to journalism is reviewing, & of this I do a great deal. My facility in it steadily increases, & so, I think, does the quality of my articles. I have a book of criticism all ready for printing, & it ought to make a bit of a stir; but Grant Richards has had it for months, & neither I nor my agent can get a definite answer out of him. I am on the point of getting angry at the swine.
>
> (10 February 1901, in Bennett 1968: 145)

Bennett was obviously productive but this productivity was of a particular kind, predominantly reviewing and literary criticism. A little later in the same year H.G. Wells wrote enclosing a copy of his non-fiction work *Anticipations,* with the friendly request that 'if it takes you [. . .] do something to propagate my gospel' (25 November 1901, in Bennett and Wells 1960: 25). This suggests Bennett's reputation in the journalistic field, but as Bennett's reply makes clear, he was in need of help to penetrate the literary field. In his reply, Bennett reminded Wells of his promise to introduce Bennett to his literary agent, J. B. Pinker (13 December 1901, in Bennett and Wells 1960). This Wells did, and from the close of 1901 Pinker began to act on Bennett's behalf. Interestingly, one of Pinker's first recommendations was that Bennett continue to endorse Wells' work, doing so in an article for *Cosmopolitan Magazine*. The strategy was seemingly one of establishing reputation by association and of writing one's way from the world of journalism into the world of literature. Bennett's choice of publisher also reflects his status on the fringes of the literary world. Grant Richards was a relatively new publishing house (it had only been established in 1897) and one that tended to be favoured by new and rising names. Few writers stayed with the publisher once they had established their literary reputation, preferring to move on to the likes of Chapman and Hall and Fisher Unwin (Brockman 2004). During his time at Trinity Farm, then, we see Bennett gradually writing his way into more significant literary circles, or closing the gap between his world and that of his literary peers. Yet, as his letter to Sturt and his early commission from Pinker remind us, Bennett's reputation needed to be built through apprenticeship both in cognate fields and to more experienced authors.

Arranging the interior: furniture and bibelots

Bennett's move to Trinity Farm and his insistence on a particular decorative order was about generating a specific affect: a vision of who Arnold Bennett the author was. This was not a process unique to Trinity Farm; each of Bennett's homes underwent a process of internal manipulation that was expressive of his subjectivity. If we return to Les Néfliers, Bennett's marital home in France, the textual register of his decoration is supplemented by a visual one that reveals something more about his process of interior arrangement and the affects he sought to generate. Recent work on the geographies of architectural practice sees the process of design and inhabitation as something more than the search for homeliness or belonging; rather, it examines place-making, or home-making, as the production of certain forms of affect (Jacobs and Merriman 2011; Kraftl and Adey 2008). Affect is 'a sense of push in the world' (Thrift 2004: 64), something that happens to 'catch people up in something that feels like *some*thing' (Stewart 2007: 2). As such, affect emerges from the relations between different bodies rather than being immanent within any one person or thing. There is a possible tension here, however, between the way in which interiors are shaped by one's identity while simultaneously affecting this sense of identity. It is a tension that begins to dissipate when we understand interiors as always in process; we may take days out to decorate and design them but interiors are always evolving as stuff accumulates, things are rearranged and colours and patterns fade with the passage of time. At the same time, how one is caught up in the interior will itself vary over the duration of inhabitation. On arrival one may be overwhelmed by the place and it is only over time that one begins to exert oneself on the interior. The interior, then, is always a canvas in the working out of identity; its significance is never fixed for it is always being refreshed and reassembled, allowing it to be affective and affected.

The canvas of a writer's room, Fuss (2004) observes, is often replete with material things – books, artefacts, technologies and mementoes – that catch up the self in reverie, memory, fantasy and daydream, and animate intellectual labour. If we peer into Bennett's writing room at Les Néfliers we find a style that a friend had earlier described as: 'Empire, verging on the Louis-Philippe. His taste struck me as being very good, but erring on the side of reticence. *Now* I should love it; one gets less flamboyant as one gets older' (Joll n.d.). Bennett's writing room (Figure 1.1, page 15) houses a desk in the Empire style and this is juxtaposed with a chair that in its curved and softer form is more in the Louis-Philippe style. Alongside this, the room is notable for its pale colour scheme with the only hint of colour coming from the suggestive edge of the curtains that are evident on the left-hand of the photograph. In France at this time, furniture was intimately bound up with ideas of taste and social power (Auslander 1996). Where it had previously been indicative of political power, the growth of mass production and consumption in the later nineteenth and early twentieth centuries had transformed furniture in to a signifier of social status and good taste. An interior was moulded to convey and catch up selves in the perpetuation and circulation of social capital. It was not just pieces of furniture and their provenance that were important, though; so too was their spatial arrangement. Interior design manuals and the work of aesthetes like the de Goncourt brothers, for instance, were strong influences on the fashion for stylistically harmonious and historically cohesive rooms (Lasc 2013; Rich 2003). There is evidence of this in other rooms within Bennett's home, which contained a salon very much in keeping with the striking dignity of the Empire style (Figure 1.2, page 16). Thus, the historical and stylistic pastiche within Bennett's theatre of composition seems a little out of keeping.

The pastiche of Bennett's writing room may well be expressive of his love for routine. As Fredrick Marriott observed, Bennett would always follow the same route on his morning walk in order to lessen the risk of distractions (Marriott n.d.) and the same idea may well apply to the interior of his theatre of composition. An interior simple in its layout and organisation, and one that emphasises comfort rather than cohesion, precludes sensory excitement and closes down opportunities for daydreaming and fantasising, allowing Bennett to focus on his writing. This may be one reason why the desk is against the wall rather than in front of the window, for a window view could potentially distract Bennett from his purpose of writing. The austere layout and design of the room is, however, convoluted somewhat when we look to Bennett's biblioteering[3]. The bibelots, or knick-knacks and decorative objects that adorn the room, such as the pictures on the wall, the vase of flowers just visible over Bennett's left shoulder, the letters and papers scattered across the desk and the small bookcase on the desk's edge, all suggest distractions. Studies are often considered the natural home of collections, clutter and material gatherings, facilitating free associations and a creative thought process (Fuss 2004; Rice 2007; Sjöholm 2013); yet this grates somewhat with Bennett's predilection for routine and order. However, Bennett's bibelots are not simply assembled at random. The pictures, for instance, are arranged symmetrically; at the centre there is a large landscape painting flanked on either side by two smaller pictures, the upper one slightly larger than the lower one, which is, on both sides, a head and shoulders portrait image. At the bottom and seemingly unifying the two sides is a smaller picture on a landscape orientation. The whole effect draws the eye downwards towards the desk. Within the context of the pictorial arrangements in the home of the de Goncourt brothers, Pamela Warner (2008) argues that such symmetry was intentional, serving to foster aesthetic contemplation, to catch one up in a mental something or somewhere, by focusing and contracting the viewer's field of vision. This gives the bibelot a somewhat different inflection, for rather than operating as a distraction that takes the mind off in new directions, it focalises and concentrates artistic intentions. Bennett was greatly influenced by the artistic ideals of the de Goncourt brothers, modelling his journal on theirs and admiring their naturalistic style (Drabble 1975). It is not unlikely, therefore, that their influence crept into Bennett's approach to interior design and the manner in which this was integral to the building and expression of his authorial identity.

Figure 1.1 Bennett's writing room, Villa des Néfliers.

Image courtesy of Keele University Library, Arnold Bennett Papers, ABK 35.

Figure 1.2 The salon, Villa des Néfliers.
Image courtesy of Keele University Library, Arnold Bennett Papers, ABK 35.

Importantly though, Bennett's preoccupation with interior design was not about generating a public identity *per se*; rather, it serves as a means of catching the self and turning it from the world outside to the world inside. This returns us, then, to where we started, for in arranging his interior Bennett appears to be trying to create and secure a mental and material space, or black-box, which generates particular affects. Things were arranged within to foster productivity both on the page and in place: through his day-to-day compositional practices which, in turn, contributed to the fashioning of his reputation abroad. Tracing Bennett's interior arrangements allows us insight into some of the practices and processes that go into the making, or *catching*, of the interior as a black-box: a space few literary scholars choose to enter. What it reveals is that the practices of assembling and arranging transgress spatial confines; Bennett may well have been keen to order the material space of his study, but just a cursory peek into this room demonstrates that the things that wind up here have ideological genealogies and international trajectories that make the walls of this writing space always, and necessarily, permeable.

Conclusions

Thus far, this chapter has said relatively little about Bennett's represented spaces: the interior spaces he so painstakingly detailed in *Clayhanger* and *Hilda Lessways*, and which have, partly as a result of Virginia Woolf's critique, led in some ways to his marginalisation within literary scholarship. It has, instead, focused on what Henri Lefebvre (1991) would term his lived spaces; the sites of his spatial practices, of which writing was one. In concluding, though, it is worth saying something more about the relationality between the lived spaces and the represented spaces – for Bennett was an inhabitant of both.

It is via a detour to Mikhail Bakhtin's work on the chronotope, the representation of space–time within the narrative world, that we come at this relationality. According to James Kneale (2011), Bakhtin saw public spaces – streets, market places, doorways and so forth – as the transformative, or threshold spaces within a novel. It was in these spaces that things happened, action moved on and events unfolded. In contrast, interior spaces were inert; they held-up and slowed the action, functioning more to fill than unfurl time. If we return to Les Néfliers, to Trinity Farm and even to the brief visit we paid to Bennett's first flat in Paris, the transformative spaces of his life, and the spaces he registered and wrote about quite extensively in his journals and letters, were these lived, interior spaces. It was here that he tried to forge his literary credentials, his authorial identity and to fashion spaces in which he could work productively. That spaces such as these were so integral to Bennett's own life should make us re-evaluate Woolf's accusation of detailism in relation to the representation of these spaces within his fiction. Take, for instance, the London lodgings that Richard Larch occupies in *A Man from the North*. These are presented to us in detail;

> [. . .] a long, rather low room, its length cut by the two windows [. . .] between the windows a table with a faded green cloth, and a small bed opposite; behind the door an artfully concealed washstand; the mantelpiece painted mustard yellow, bore divers squat earthenware figures, and was surmounted by an oblong mirror framed in rosewood [. . .] The walls were decorated with a pattern of giant pink roses.
>
> ([1898] n.d: 4–5)

The effect is to make the reader tangibly aware of the solidity of the room. More than this though, the representation of these interior spaces was bound up with the development of character. The interior of Larch's room was not a way of gaining access to his consciousness but it was a way of measuring his self-development; the interiors he inhabited were fulcrums of change and energy within both plot and character. As the novel moves on we encounter Larch through a series of interiors which, as Robert Squillace observes, 'create different perspectival contexts' in which the character develops and which 'produces very different impressions' (1997: 25). Through this prism, Bennett's interior detailism is a way of allowing us access to the plurality and peculiarities of character: to the inconsistencies of self in time and space. As the novel closes we find Larch outside not inside a house. With his literary and romantic ambitions spent, being on the wrong side of the door suggests the limit of character development: Larch can go no further. In going into his house he turns his back on the world outside.

Let us close by folding this relationship back on itself once more and returning to the chapter's point of departure: that composition takes place within material locations and the objects that crowd these places. The manner in which interior spaces are decorated and the nature of their organisation and orientation are critical influences on how writing happens, is prevented from happening or is envisaged as happening. This chapter, however, has demonstrated that the interior space of composition is something more than the three-dimensional space enclosed by the walls of the study. Instead, the space of composition pulses with artistic connections that push beyond the confines of these walls to encompass the house's other inhabitants, both visible and invisible, the house's relationships to broader artistic circles, and the ties between the lived and represented spaces of the house. The nature of these pulsations is not always straightforwardly artistic, though. Being a writer and doing writing are two different things, as is writing as a relative novice vis-à-vis writing as an established and recognised author. Thus, interior concerns, both material and psychological, are contingent upon time and place; how the interior matters to writing – and to character – must attend to the span of an authorial career and to its unique patterning of everyday life.

Notes

1 Identity-based political aesthetic is used to register the difference between the often explicit social politics that tend to be associated with the novels of Arnold Bennett, H.G. Wells and John Galsworthy and the subtler reflections on identity, particularly gender and sexuality, which haunted modernist fiction.

2 The term 'power geometry' was coined by Doreen Massey in *Space, Place and Gender* to explain how different social groups were placed and affected differently by the flows and mobilities of globalisation. Some, often the more affluent, were able to move unimpeded and at speed around and through the world. In contrast, the poorer within the world were often isolated by their inability to access communication technology and high speed travel.

3 Bibelot is a French term for knick-knack, collectible and curiosity. During the nineteenth century the word came to encompass the wider practices of collecting, classifying and describing. Bibelots became a popular trope within French prose writing during the latter half of the nineteenth century, in part to help express the nature of character, but also as expressions of the growth of material culture itself, as a result of industrial manufacture and mass consumption. See Janell Watson (1999) for a detailed discussion of bibelots and bibeloteering in relation to literature.

References

Anderson, A. 2013. Drawing rooms: a backward glance – Fashioning an individual drawing room. In G. Downey, ed. *Domestic Interiors: Representing Homes from the Victorians to the Moderns*. London: Bloomsbury, pp. 39–6

Anderson, B., Kearnes, M., McFarlane, C. and Swanton, D. 2012. On assemblages and geography. *Dialogues in Human Geography*, 2 (2), pp. 171–189.

Anderson, B. and McFarlane, C. 2011. Assemblage and geography. *Area*, 43 (2), pp. 124–127.

Ardis, A. 2002. *Modernism and Cultural Conflict, 1880–1920*. Cambridge: Cambridge University Press.

Ash, J. 2009. Emerging spatialities of the screen: video games and the reconfiguration of spatial awareness. *Environment and Planning A*, 41 (9), pp. 2105–2124.

Auslander, L. 1996. *Taste and Power: Furnishing Modern France*. London: University of California Press.

Barnes, T. 2004. Placing ideas: genius loci, heterotopia and geography's quantitative revolution. *Progress in Human Geography*, 28 (5), pp. 565–595.

Barthes, R. 1977. *Image, Music and Text*. New York: Hill and Wang. Trans. S. Heath.

Benjamin, W. [1930] 1999. *The Arcades Project*. Cambridge MA: Harvard Bellknap.

Bennett, A. 1968. *Letters of Arnold Bennett, Vol. II, 1889–1915*. J Hepburn, ed. London: Oxford University Press.

Bennett, A. 1932. *The Journals of Arnold Bennett, Vol. I: 1896–1910*. N. Flower, ed. London: Cassel & Co.

Bennett, A. 1929. Books and Persons. *The Evening Standard* (28 November 1929).

Bennett, A. 1914. *The Author's Craft*. London: Hodder and Stoughton.

Bennett, A. [1911] 2008. *Hilda Lessways*. Looe: Stratus Books.

Bennett, A. [1910] 2000. *Clayhanger*. London: Penguin.

Bennett, A. [1908] 2007. *The Old Wives' Tale*. London: Penguin.

Bennett, A. [1898] n.d. *A Man from the North*. Leek: Churnet Valley Books.

Bennett, A. and Wells, H.G. 1960. *Arnold Bennett and H.G. Wells: A Record of a Personal and Literary Friendship*. H. Wilson, ed. Urbana: University of Illinois Press.

Bennett, M. 1925. *Arnold Bennett*. London: A.M. Philpot Ltd.

Bourdieu, P. 1993. *Field of Cultural Production: Essays on Art and Literature*. Oxford: Blackwell.

Bourriaud, N. 2002. *Relational Aesthetics*. Dijon: Du press du reel.

Brace, C. and Johns-Putra, A. 2010. Recovering inspiration in the spaces of creative writing. *Transactions of the Institute of British Geographers*, 35 (3), pp. 399–413.

Brockman, W.S. 2004. Richards (Frankin Thomas) Grant, 1872–1948. In E. Goldman, ed. *Oxford Dictionary of National Biography*. Oxford: Oxford University Press.

Castle, G. 2015. Bootmakers and watchmakers: Wells, Bennett, Galsworthy, Woolf and modernist fiction. In G. Castle, ed. *A History of the Modernist Novel*. New York: Cambridge University Press, pp. 137–152.

Darby, H.C. 1948. The regional geography of Thomas Hardy's Wessex. *The Geographical Review*, 38 (3), pp. 426–443.

Drabble, M. 1975. *Arnold Bennett*. London: Omega.

E.B.S. 1896. Some recent designs by Mr. C.F.A. Voysay. *The Studio*, 38 (Feb), pp. 209–218.

Farrell, M. 2001. *Collaborative Circles: Friendship Dynamics and Creative Work*. London: University of Chicago Press.

Fuss, D. 2004. *The Sense of an Interior: Four Rooms and the Writers that Shaped them*. London: Routledge.

Gieryn, T.F. 2002. What buildings do. *Theory and Society*, 31 (1), pp. 35–74.

Gilbert, E.W. 1960. The idea of the region: Herbertson Memorial Lecture. *Geography*, 45, pp. 157–175.

Hones, S. 2014. *Literary Geographies: Narrative Space in 'Let the Great World Spin'*. Basingstoke: Palgrave Macmillan.

Jackson, L. 2007. *Twentieth-century Pattern Design*. New York: Princeton Architectural Press.

Jacobs, J. and Merriman, P. 2011. Practising architectures. *Social and Cultural Geography*, 12 (3), pp. 211–222.

Joll, E. no date. *The Young Greens*, unpublished manuscript. Stoke: Potteries Museum and Art Gallery.

Kneale, J. 2011. Plots: space, conspiracy, and contingency in William Gibson's Pattern Recognition and Spook Country. *Environment and Planning D: Society and Space*, 29, pp. 169–186.

Kraftl, P. and Adey, P. 2008. Architecture/affect/inhabitation: geographies of being in building. *Annals of the Association of American Geographers*, 98 (1), pp. 213–231.
Lasc, A. 2013. Interior decorating in the age of historicism: popular design manuals and the pattern books of Édouard Bajot. *Journal of Design History*, 26 (1), pp. 1–24.
Latour, B. 1987. *Science in Action: How to Follow Scientists and Engineers through Society*. Cambridge MA: Harvard University Press.
Latimer, C. 1988. The division of the wall: the use of wallpapers in decorative schemes, 1870–1910. *The Journal of the Decorative Arts Society*, 12, pp. 18–25.
Lefebvre, H. 1991. *The Production of Space*. Oxford: Blackwell.
Lehan, R. 2005. *Realism and Naturalism: The Novel in an Age of Transition.* Madison: The University of Wisconsin Press.
Livingstone, D. 2003. *Putting Science in its Place: Geographies of Scientific Knowledge*. London: University of Chicago Press.
Lorimer, H. and Spedding, N. 2005. Locating field science: a geographical family expedition to Glen Roy, Scotland. *British Journal of the History of Science*, 38 (1), pp. 13–33.
Marriott, F. no date. *Adventures with Arnold Bennett*. AB Misc 2: Keele University Special Collections and Archives.
Massey, D. 1994. *Space, Place and Gender*. Minneapolis: University of Minnesota Press.
Massey, D. 2005. *For Space*. Oxford: Sage.
Miller, D. 2008. *The Comfort of Things*. Cambridge: Polity Press.
Paxton, N. 2000. Eclipsed by modernism. In L. Hapgood and N. Paxton, eds. *Outside Modernism: In Pursuit of the English Novel, 1900–1930.* Basingstoke: Macmillan, pp. 3–21.
Pevsner, N. [1936] 1991. *Pioneers of Modern Design from William Morris to Walter Gropius*. Harmondsworth: Penguin.
Rice, C. 2007. *The Emergence of the Interior: Architecture, Modernity, Domesticity*. London: Routledge.
Rich, R. 2003. Designing the dinner party: advice on dining and décor in London and Paris, 1860–1914. *Journal of Design History*, 16 (1), pp. 49–61.
Saunders, A. 2010. Literary geography: reforging the connection. *Progress in Human Geography*, 34 (4), pp. 436–452.
Sjöholm, J. 2013. The role of the art studio in contemporary artistic production. Centre for Research on Innovation and Industrial Dynamics, Research Paper 2013, pp. 1–34.
Squillace, R. 1997. *Modernism, Modernity and Arnold Bennett*. London: Bucknell University Press.
Stewart, K. 2007. *Ordinary Affects.* Durham NC: Duke University Press.
Thrift, N. 2004. Intensities of feeling: towards a spatial politics of affect. *Geografiska Annaler, Series B Human Geography*, 86 (1), pp. 57–78.
Vadillo, A.P. 2007. Aestheticism 'at-home' in London: A. Mary, F. Robinson and the Aesthetic Sect. In G. Cunningham and S. Barber, eds. *London Eyes: Reflections in Text and Image*. Oxford: Berghan Books, pp. 59–78.
Vallance, A. 1892. The Furnishing and Decoration of the House. *Art Journal*, 53, pp. 44–49.
Warner, P. 2008. Framing, symmetry and contrast in Edmund de Goncourt's aesthetic interior. *Studies in the Decorative Arts*, 15 (2), pp. 36–64.
Watson, J. 1999. *Literature and Material Culture from Balzac to Proust*. Cambridge: Cambridge University Press.
Wimsatt, W.K. and Beardsley, M.C. 1946. The Intentional Fallacy. *Sewanee Review*, 54(3), pp. 464–488.
Woolf, V. [1924] 1959. *The Captain's Death Bed and other Essays*. New York: Harcourt Brace Janovich.

2 Holland Park, west Kensington, London

When the artist and sculptor George Frederic Watts died in 1904 he left behind not only an aesthetic heritage but one of London's most famous artistic neighbourhoods. Watts, along with his fellow artist and sculptor Frederic Leighton, who was president of the Royal Academy from 1878 until his death in 1896, had been two of the principle forces in what has subsequently become known as the Holland Park circle (Drakes 1999). The Holland Park circle, according to Caroline Drakes, had its origins in the 1850s when Watts helped Henry and Sara Prinsep take out a lease on Little Holland House, the dowager house on the Holland Park estate in west Kensington. For 21 years Little Holland House became a hub for London's literati. A regular feature was the Sunday 'at homes' that saw gatherings of the city's artists and writers that included Edward Burne-Jones, Gabriel Rossetti, George du Maurier, Alfred Tennyson, William Burges, George Millais, and Elizabeth Barrett Browning. Leighton moved to the area in 1864 when he acquired a plot of land on the Holland Park estate. On this he built 2 Holland Park Road: his palace to art. As the century progressed this house was to become another fulcrum within the Holland Park circle, playing host to musical soirees and dinner parties and a guest list that included the Prince of Wales (Drakes 1999). By the 1880s the combined fame of Watt and Leighton had established Holland Park as an artistic enclave. Many of the early visitors to Little Holland House had subsequently bought their own properties on land leased from the Holland Park estate. By the 1890s Melbury Road, which cut across the top of Holland Park Road was home to artists and designers that included William Burges, Luke Fildes, Marcus Stone and Colin Hunter.

Today we might term this a creative cluster, a gathering of individuals that sees in spatial proximity opportunities for information exchange, professional advancement and modish or niche lifestyles (Scott 2006; Florida 2002). It is not just spatial proximity though, that is important to cultural production, so too is the particularity of the place as a lived and inhabited social world. The artists who came to Holland Park came to be part of something and they were attracted by something. That something may well have been reputation-by-association, the need for intellectual belonging or the capacity to practice and pursue a particular lifestyle, yet it is likely that much of their daily life was spent in far less rarefied pursuits. The particularity or specialness of place, as Doreen Massey observes,

' . . . is not some romance of a pre-given collective identity . . . [r]ather, what is special about place is precisely that throwntogetherness, the unavoidable challenge of negotiating a here and now' (2005: 140). This is not to suggest that collective identity is unimportant, but that this identity is not a transcendent reality, or a given and settled coherence; instead, it is a contingent and ongoing making that emerges from the everydayness of the world. A sense of this is captured in Sara Blair's (2004) study of the particularity of fin-de-siècle Bloomsbury, which explores this area of London as a lived literary world and not merely an imagined literary project. It is, Blair argues, a place tied up in progressive politics, the radical suffragist movement, class dissent and transnational pressures, and these forces and ways of life are entangled with the cultural production *in* and *of* the area. Bloomsbury both lends itself to and invites this kind of study, for it is, of all London's neighbourhoods, the one most popularly and frequently understood as a cultural phenomenon before it is a lived world. Yet, as the world that gathered within and around Holland Park, west Kensington demonstrates, Bloomsbury while the most well-known cultural phenomenon today, was not the only one in the early decades of the twentieth century.

The cultural production in and of Holland Park appears, first and foremost, as one dominated by a community of visual artists whose most formative years ran from the 1850s to the 1890s. After the death of Leighton in 1896 and Watt's removal to Guildford in 1891 it is argued that the circle began to unfurl and the area lost its artistic pull and momentum (Drakes 1999). However, cultural and demographic change is never immediate and the area continued to attract artists throughout the early years of the new century. William Holman Hunt for instance, did not move into Melbury Road until 1903, and it was not just visual artists who were attracted to the area. In the 1850s and 1860s the salons at Little Holland House had attracted writers such as W.M. Thackeray, Charles Dickens and Elizabeth Barrett Browning, and as the century turned a new wave of writers began to move into the area. In 1907 May Sinclair moved to Edwardes Square, a year later Ford Madox Ford took out a lease on a flat at 84 Holland Park Avenue and began *The English Review* from these lodgings, in the same year Violet Hunt moved to South Lodge on Campden Hill, while the 'rebel writer' Evelyn Sharp occupied lodgings in Dukes Lane and the playwright Laurence Housman lived in Pembroke Gardens. Another incumbent was John Galsworthy who, in the autumn of 1905 moved with his new wife, Ada, to number 14 Addison Road. The writers who congregated in this area in the early years of the twentieth century have never been considered as part of the Holland Park circle, not least because they came too late and worked with letters not canvases, but as with Bloomsbury it was inescapably a place of cultural production at this time. Drawing on Massey's idea of the 'throwntogetherness' of place and Blair's urging that we read literary places as lived worlds not cultural phenomena, this chapter tunes in to Galsworthy's material social world – the happenstance conjunctures and the networks of relationship, both spatial and temporal – of Holland Park in the first decade of the twentieth century. Such an approach allows us, as Blair observes, to understand literary writing as 'a determined response to the specific space in which it takes shape, advertises its cultural value and contests for social power' (2004: 814). The intention is not to flag influence from lived experience, but rather, to provide a rich portrait of place, what Thrift (1996) terms, the 'productive locale' in which we live our lives.

The social world of Kensington: 14 Addison Road

One of the vectors that Blair uses to explore the lived social world of Bloomsbury is the guides and maps, designed to make London known spatially, culturally and economically, that proliferated in the later nineteenth and early twentieth centuries. Of these, *Baedeker's London and its Environs* (1901) and *Ward Lock's Pictorial and Descriptive Guide to London and its Environs* (1908) were two of the most popular tourist guides to the city. What is interesting about both is the peripherality that is accorded to the neighbourhood of Holland Park. Discursively, it is subsumed within the larger neighbourhood of Kensington, which is bookended by Kensington Palace to the east and Holland House to the west, and for both guides the area between is notable for its desirable residences and its past association with such personalities as Thackeray, Leighton and Macauley. From Holland House, *Ward Lock* moves back along Kensington High Street to South Kensington, while *Baedeker* moves north into Bayswater. The area west of Holland Park, where Addison Road lies is very symbolically off-the-map, both discursively and pictorially. When the Galsworthys moved to Addison Road in the autumn of 1905 perhaps they sought this peripherality. Earlier in the year, after Ada's divorce, Galsworthy, in an effort to pre-empt social disapprobation, had renounced all his club memberships and company directorships preparatory to marrying Ada. As James Gindin (1987) narrates, even at this time this was a little excessive, but Galsworthy's sense of social etiquette was strong. Several decades earlier Holland Park's charm was that it was 'not, like South Kensington and Belgravia, a mere hanger-on and copy of Mayfair, but, like Chelsea, it ha[d] a life and history of its own. The inhabitants . . . [were] mostly working people, less trammelled by the rules of fashion than they are a few hundred yards further east, and obeying in many respects a peculiar code of their own (Lang 1886: 26–7). No doubt, the area had changed by 1905, most notably in the death or removal of its two most prominent artists, yet it continued to be a cultural hub and may well have offered the Galsworthys a home that escaped some of the pressures that accrued within the city's social, political and commercial centres. They were not tucked away but nor were they in a place where they were obviously curiosities or an embarrassment; they were in a London beyond the popular gaze.

That said, if we pursue another of Blair's spatial vectors, namely Charles Booth's poverty maps that were generated during the later decades of the nineteenth century, the neighbourhood acquires a different hue. Where Bloomsbury was dominated by pink and red, indicating the relative security of the lower and more comfortable middle classes, the area around Holland Park was even better-off. Holland Park itself was yellow, the colour of the upper echelons of the middle classes and the very wealthy, and so too were the streets that snaked around its boundary: Holland Park to the north, Phillimore Gardens to the east and Melbury and Addison Roads to the west stood sentry-like around the Park guarding against an encroaching sea of red. These streets grew up in the early nineteenth century, on land belonging to the Holland Park estate. They developed in a piecemeal fashion, as parcels of land of varying sizes were sold off to a number of speculative builders (Sheppard 1973). This method of development was indicative of the changing financial demands placed upon the Fox family (incumbents of Holland House), but it was also intended to encourage particular kinds of socio-spatial development: detached villas, suburban mansions and charming town houses rather than terraced housing, and architecturally diverse rather than uniform developments. Every town or architectural plan, Gieryn (2002) argues, is a blueprint for social life, which once translated into material form stabilises social action and perpetuates a particular social imaginary. Hence, if Bloomsbury with its runs of Georgian terraces, each very similar to the next, was a world of bankers, lawyers and the salaried professions, where status and worth were measured in relation to one's neighbours, it would seem that this corner of west Kensington, with its stress on variety and inimitability, was intended to foster a more bohemian and non-conformist way of life.

Walking down Addison Road one is struck by the hotchpotch nature of the housing. Number 14 Addison Road, the house the Galsworthys would move into in 1905, and its semi-detached neighbour at 15, are relatively small in comparison to some of the other houses in the road. Buffeted from the street by sizable front gardens, they have cast iron balconies and wood trellis verandas giving them an almost cottage-like appearance. Across the road are more sizable, detached dwellings often neo-gothic in style while further down, the two-storey style of numbers 14 and 15 give way to three-story town houses, some in short-runs of terraces. Walk round the corner into Melbury Road and thence in to Holland Park Road and these houses pale in comparison to the imposing immensity of red-brick mansions and townhouses, often in Gothic or Flemish style and replete with such flourishes as cupolas, porticos, mullioned windows and tall chimneys, which dominate these two streets. These were frequently designed by artists who had bought land on preferential terms from the Fox family, and who saw their houses as another expression of their art and for whom large, private studios were an integral necessity within their homes. Buildings do not stay the same forever, but, as Gieryn (2002) argues, they accumulate social momentum and currency that can take years to dispel or dilute. Thus even in 1905, when the Holland Park circle was on the wane, the area remained, as we shall see, progressive and a little offbeat.

Before exploring two aspects of this progressiveness though, something more remains to be said about the immediate social world of Addison Road in the first decade of the new century. Façades only take us so far, what we read off the built landscape may be at odds with what goes on behind a street's front doors. This was one of the problems with Booth's poverty maps, which were premised on appearance and aesthetic not people. An area may have an artistic feel or give the impression of wealth, but these are rarely all-encompassing atmospheres. Thus, a more productive and nuanced version of the lived social world of Addison Road emerges in the 1901 census, enabling us to move beyond the street's public face and explore the homes and people that comprise a neighbourhood: the very immediate locale of Galsworthy's cultural production (British Census 1901).

A social snapshot: the space and time of Addison Road in 1901

In 1901 number 14 Addison Road was inhabited by Delissa Joseph, an architect, and his wife, Lily. Joseph had already designed several buildings in the city of London by 1901, as well as a number of mansion flats and apartment buildings that played such a central role in altering London's residential fabric at this time (Dennis 2008). A close neighbour was Frederick Coles, a stockbroker, and his wife Mabel who resided at number 12, a slightly more imposing house of several storeys. Coles was the son of the self-made businessman John Coles, who left a sizable fortune on his death in 1919. Further down, at number 16, was another stockbroker, the Italian Piero Silo, and his American wife Jeannie. Although in several of Trollope's novels, most notably *The Way We Live Now* (1875), banking was often seen as an unsavoury and ungentlemanly profession, by the turn of twentieth century a career in banking and brokerage was far more socially acceptable and access often depended upon blood, marriage and education. Thus, it was a relatively closed world with strong links to the landed aristocracy, but it was not so tightly sealed so as to preclude all who fell outside this purview by default of birth or connection. Many of London's merchant banks and stockbroking companies had been established by and continued to employ significant numbers of foreigners, meaning that inclusion, for some, often depended upon cultural difference not sameness (Michie 1998). The area around Holland Park however was not a popular one with powerful banking families, for whom Mayfair and Belgravia were far more attractive, underscoring some of the spatial and social cleavages within the city's financial community (Cassis 1985).

Another measure of the area's transnationalism was Phillip Mennell who lived with his wife Ellen, their three children and two domestic servants a little further along the street at number 8. In the 1870s Mennell had migrated to Australia working first as a journalist before buying shares in a newspaper company. On his return to Britain in the 1880s he took over the reins of the *British Australian and New Zealand Mail*. A hub for the Australian community in London, under Mennell's proprietorship the newspaper moved away from its early focus on finance to reflect the growing diversity of Australians and Australian-interests in London. Moreover, it became a vocal supporter of British migration to the colonies (Sleight 2009). Not all the houses within the street had a professional income, a number were home to individuals of 'independent means' like Mary Mivart who lived at number 17 and the widow, Malvina Mosely, and her daughter Julia who lived at number 13. Although it is difficult to decipher how large their means were, it is unlikely that they were extensive. Martin Mosely for instance, also lived at number 13 with his mother and sister, yet unlike them he had no independent income and was employed as a banker's clerk. A final illustration of the area's complexity

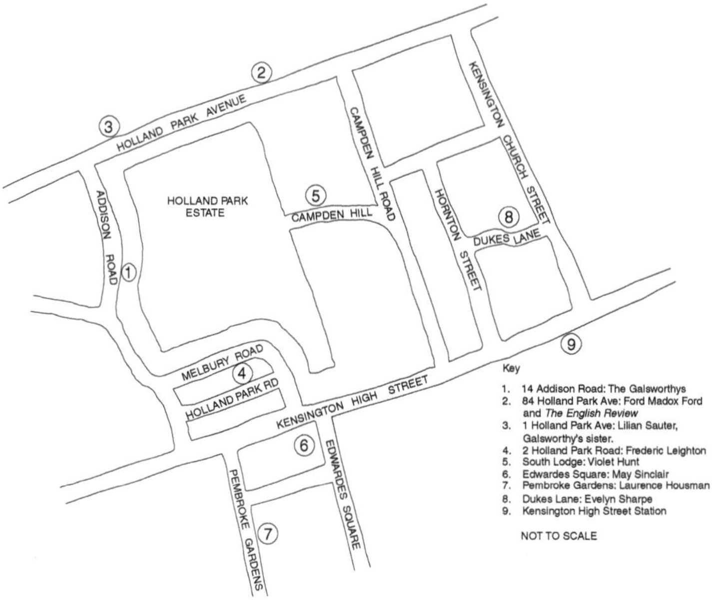

Figure 2.1 The neighbourhood of Holland Park, west Kensington.

lay just round the corner in Holland Park Road where Frederick Leighton's house had been opened as a museum to his life and work.

The immediate world the Galsworthys would move into in 1905 then, was solidly professional. In the main incomes were earned and wealth accrued through professional rank or seniority. There was evidently some inherited wealth but this seems to have been predominantly of a comfortable rather than huge degree. A contrary picture arises through if we look to the area's domestic servants. Booth designated streets yellow where there were three or more servants and in most cases the houses in Addison Road had two if not three or more servants. Indeed, some of the houses were made up solely of servants, such as number 15 where Emma Crews, cook, and Minnie Chapple, housemaid, resided, or Addison Lodge that comprised six staff including footmen and a secretary but no head of household. This may indicate seasonal residences or the commandeering of smaller properties purely to home domestic help for the street's larger residences. The number of domestic servants vis-à-vis owners gives substance to the road's privileged social world, but it also points to a road of two worlds: the one wealthy the other far less so, the comfort of one dependent upon the reproductive labour of the other and the relative emancipation and independence of the road's wealthy women contingent on the domestic assistance of others (Light 2007; Todd 2009).

The social world(s) of the street are further complicated if we move beyond the front door of number 21 and enter the boarding house that lay beyond. Boarding houses performed an important bridging function within turn of the century

London, particularly for women. Not only did they offer respectability and security for the growing number of independent women, but, as Angela Woollacott (2001) argues, they facilitated female mobility and assimilation; they offered affordable accommodation and an introduction to an immediate social world for the growing number of itinerant women who came not just from the provinces but from the colonies too. A sense of this is conveyed by the boarding house at number 21, which was run by Mary Gregory, who, along with her three domestic servants, Annie, Ellen and Rachel, looked after four visitors: Eliza, Ada, Marion and her two-year-old daughter Doris. In 1881 the house had been owned by Mary's father, a man of independent means, and it is likely that in order to continue to afford the upkeep of the house Mary had to take in lodgers. This calls attention to the nature of inhabitation within this place, intimating that 'being-in' this house was an affectively different experience for its inhabitants (Jacobs and Merriman 2011); for Mary it had an emotional history and resonance, for the lodgers it likely had a more utilitarian significance both spatially and temporally, while for those who worked within it, for Annie, Ellen and Rachel, it may well have been a means to a different domestic end. At the same time, as owners, lodgers and workers their practical experiences of this house, their ability to move within it, to permeate its depth and to invite guests in was, for most, somewhat curtailed. From the 1911 census it is known that this house had 12 rooms inclusive of the kitchen, but excluding bathroom, lobby and scullery. At most, it is likely that each lodger had two rooms while the servants likely had to share accommodation. Thus, what Lloyd Jenkins (2002) terms the permeability of this house both to its inhabitants but also to the street at large was not straightforward; rather than being the threshold to a home the front door was the threshold to multiple, differently positioned and differently accessed homes. Moreover, for those who worked there it was likely that it was not a threshold at all, but a symbol of wealth and position.

The nature of inhabitation may also owe to the location of 21 Addison Road, for rather than facing on to the street it gazed on to the side of number 17 with the Holland Park estate falling in to its peripheral vision. It was part of a small terrace of Georgian houses tucked away down a private driveway that ran perpendicular to Addison Road and was surrounded by a tall boundary wall. Its position suggests a heightened sense of privacy; there were fewer opportunities to get caught up, especially visually, in the world beyond the house's four walls. Although enclosed in a physical sense, the trajectories of the house's inhabitants speak of a much wider social world. Marion, originally from Ireland, had recently returned from India where her daughter had been born, while Eliza had moved to London from Jamaica. Neither Marion nor Eliza worked suggesting that they had a private income or were, perhaps, recent arrivals from abroad. Ada's world in contrast seems quite tightly drawn, born in London she worked as an art teacher. The census only tells us so much, revealing in some ways more absences than presences; for instance, Marion was married but where was her husband, in India perhaps; how long had Eliza been living in Britain; did Ada's localness conceal greater mobility? Unanswerable in many ways, at least through the census alone, but what we can intuit is an intertwining of mobile lives. The obvious movements of Marion and Eliza, tempered by the invisibility of their daily lineaments, are augmented by the likely temporalities of Ada's daily work routine. Underpinning and facilitating these mobilities were the routines of the domestic servants, which were, for a house of this size, likely to be quite extensive comprising cooking, cleaning, errand-running and perhaps acting as ladies-maid. This is what Massey terms the event of place; the constellation of spatialities, often very mundane and unexceptional, in which there is no eternal or fixed coherence, but rather, a persistent negotiation of coexistence.

The nature of this negotiation was likely highly private, in some sense unknown or unimportant to the world outside, but in another sense it was highly visible: the goings forth of Ada, the arrival of deliveries, the running of errands and the despatching of post. It points to a street life comprised of multiple and cross-cutting temporalities. These temporalities were further enriched with the arrival of the Galsworthys in 1905, for unlike his neighbours Frederick Coles and Piero Silo, Galsworthy both lived and worked at number 14. Thus, he was not bound by the structures of a daily or weekly work routine in quite the same way and writing, which he tended to do either in the dining room or summer house and which occupied him for between 4–5 hours a day (Mottram 1956; Sauter 1967), was far from the only aspect of his literary life. The writer and poet Ralph Mottram (1956), remembers how the house became a meeting place for writers, while Rudolf Sauter recalls the Galsworthys' attendance at his parents' 'Fourth Sunday at-homes' (1967: 35) held just round the corner from Addison Road at 1 Holland Park Avenue. Thus, along with its own internal rhythms number 14 was a hub of comings-in and goings-forth, and while these two examples were not necessarily daily happenings they are communicated as regular ones, illuminating how the temporalities and spatialities of the Galsworthys were meshed with those of a wider literary and artistic community. These paths or, as Ingold (2008; 2011) terms them, lines of movement, along with those of the inhabitants of number 21, give some indication of the tapestry of time-geographies on-going within and across Addison Road, and within which Galsworthy was consciously and unconsciously immersed. If the 'at-homes' were conscious crossings of lines, a flicker of movement from the street that caught the corner of Galsworthy's eye, the lifting of his hat in lieu of salutation or a deferential nod from one of his social inferiors were the more unconscious ways in which his life was enmeshed within a generative field of possibilities.

The turn to Ingold's (2008) work allows us to consider the social world of Addison Road from a slightly different vantage point. Where Massey (2005) stresses the event of place as one of multiple relationships, Ingold argues that this is premised upon discrete and bounded entities when in fact things are made through the way they flow, mix and mingle. It is a perspective that draws from Annemarie Mol and John Law's (1994) notion of fluid space, within which there are no fixed or delimited entities but rather substances that flow, fuse, solidify, dissolve and recombine in new ways. This lacks the same kind of precision we find in Massey's work wherein the event of place happens through named and known entities, but it is a useful way of opening out the being of the writer. This, to reiterate Brace and Johns-Putra's (2010) argument, is the gamut of things a writer does other than write, but which may well shape, in often unconscious ways, the doing of their writing. Following Ingold, individuals are not one line that connects with another, but tissues of lines that become entangled with others. This presupposes a relational field not of singular meeting points but of interwoven lines: a meshwork not a network. Intermingling therefore is a condition of being and must lead us to conceptualise Galsworthy's social world not as a series of known and notable relationships that happen in fixed locations one after another, but as a meshwork of simultaneous and ongoing cominglings. These cominglings can be planned and accidental, real and imagined, remarkable and unremarkable but they generate, to borrow from John Shotter, a 'providential space' (1993: 70), from which novel possibilities and new worlds emerge.

The literary world of Kensington: 84 Holland Park Avenue

If one was looking for 84 Holland Park Avenue in the autumn of 1908 one would probably smell it before one saw it. The ground floor, opening out onto the street with an awning that bathed the interior in deep shadow, was home to a poulter's and fishmongers. Carcasses hung from the ceiling, blood and brine trickled from darkened recesses, congealing into a mat of muddy hues, and every now and then mounds of silver scales would suddenly glisten as a bloom of light flooded the doorway. Few of the inhabitants of this corner of Kensington would think 84 Holland Park Avenue anything more than a provider of fresh produce, yet if one could steer through the salty pungency of this interior to the dark stairwell behind, one would find, living and working in the three floors above, Ford Madox Ford. Ford's flat was a hive of activity at this time, for he was finalising his plans for *The English Review*: a periodical that would be integral to literature's break with all things Victorian.

Before turning to the aesthetic break that *The English Review* appeared to usher in let us pause a little longer in the unassuming location of 84 Holland Park Avenue, Kensington. The artists and writers who continued to live in this area were notable for their independent working, although they might gather at 'at-homes' or teashops like that run by Ella Abbott in Holland Street, this was a landscape of private studios and cloistered studies. It was a place of making. The place of assembly – the centre of London's industry in words, its publishing world – happened elsewhere. By 1910 London was home to 19 morning papers, ten evening papers and several hundred weekly and monthly publications (Morrisson 1996; Keating 1991). Although little has been done to disaggregate these by place, beyond alignment with the broad area of Fleet Street, a brief survey of some of the leading periodical titles of the day, such as *The Fortnightly Review*, *The Athenaeum*, *The Saturday Review*, *The Cornhill Magazine* and *The Strand*, and of more niche publications like *The Studio*, *The Nation* and *The New Age* suggests a closely drawn publishing geography. Most of these periodicals had their offices in and around Covent Garden and the Strand, with *The Fortnightly Review* at 193 Piccadilly and *The Cornhill Magazine* at 65 Cornhill marking the farthest outreaches of this print community. Kensington was a very different place to begin a new periodical; it was peripheral, suburban and very literally off-the-maps of London that were to be found in guide books to the city at this time (Lucas 1906).

The location of *The English Review* says, or perhaps reinforces, something of its aim, which was to challenge the stultifying effects of an expanded readership and commercial advertising, which together generated a periodical press orientated towards a mass market: increasingly uncritical and predominantly interested in sensation and celebrity. An office in Kensington removed one from the commercial encounters, pecuniary pressures and professional standards that may well have coalesced within the heart of the publishing world. Offices in Holland Park Avenue put one in the way of very different influences, encounters and ways of doing things. As Violet Hunt (1926) recollects, Ford maintained an open house at 84 Holland Park Avenue, albeit an openness that depended upon one's ability to tolerate the foetid smells of poultry carcasses and acceptance of the many Bolsheviks, Communists and political refugees Ford collected and gave asylum to. As a result, there was, Douglas Goldring (1943) recounts, a continuous stream of visitors to Ford's flat, both day and night, often requiring them to escape to the Shepherd's Bush Empire hall to undertake editorial work[1]. Central figures in the early months of *The English Review* were Joseph Conrad and Henry James who maintained a near continuous presence at 84 Holland Park Avenue (Hunt 1926). They, along with Ford and through a steady correspondence with H.G. Wells, were busy collecting artists for the first few issues. Many people of different literary and political hues passed through or washed up at 84 Holland Park Avenue; the physical boundaries were relatively permeable and so too were the material boundaries of the periodical. Both Violet Hunt and Wyndham Lewis recount turning up at 84 Holland Park Avenue, offering contributions and having them accepted on the spot (Hunt 1926; Ford 1999). Ford's flat became a sort of quasi-public space, part home, part business, and through which people and manuscripts flowed in a collaborative and often pragmatic enterprise.

As well as being a site of professional collaboration the flat also played host to many parties where one was likely to meet such established figures as Arnold Bennett, Hugh Walpole, Joseph Conrad, John Galsworthy, W.H. Hudson, H.G. Wells and Violet Hunt. What set these parties apart though, were the new names that were also in attendance, such as Ezra Pound, Wyndham Lewis and D.H. Lawrence. These men, Goldring (1943) observes, were the principle discoveries of *The English Review* and their presence at 84 Holland Park Avenue was indicative of the direction Ford wished the periodical to take. Ford's conception of *The English Review* took its inspiration, Mark Morrisson (1996) argues, from the *Mercure de France*, a publication designed to engage critically with a range of topics and to appeal, through its non-partisanship and disinterested inquiry, to a broad readership. A complimentary influence was Matthew Arnold's mid-Victorian vision for an international literary culture of dialogue not defensiveness (Peppis 2000). Through these lenses, *The English Review* positioned itself not as a coterie or selective publication designed to propound *a* view, but rather, one designed as a 'meeting place' of perspectives. Although it shared similarities of form and content with *The Fortnightly Review*, the cultural mainstay of the periodical press, its main difference, Morrison argues, was in its publication of unorthodox material. In the spirit of the *Mercure de France*, Ford believed that such material should be available in mainstream publications not merely in small-scale, specialist ones.

Thus, if 84 Holland Park was a physical melting pot of different literary styles and aesthetics, *The English Review* was its textual equivalent. The first issue, for instance, opened with Thomas Hardy's poem *A Sunday Morning Tragedy*, a poem *The Cornhill Magazine*, along with all the other significant periodicals, had refused to publish on the grounds of immorality. The third issue commenced with a collection of modern poetry, the first of which was *Col di Rodi* by Gerhart Hauptmann. Hauptmann was a German naturalist writer and progressive social thinker and this poem was followed by works by W.B. Yeats, Walter de la Mare, J. Marjoram[2] and finally John Galsworthy. Not only was the nature of this poetry challenging, so too was its position within the journal. A brief survey of contemporary journals reveals little poetry, and where it does feature it is rarely first, usually being preceded by prose or political commentary. Consequently, it is possible to see both place and publication as a site of experimentation: a literary laboratory. Some see this experimentation and the primacy of non-mainstream literature as indicative of the journal's modernism. Against this backdrop, Michael Schmidt (2013) has called Galsworthy's contribution dreadful and suggested that by this point *The English Review* was struggling to fulfil its aesthetic conception to publish new and radical art. In later issues the poems of Ezra Pound and D.H. Lawrence made their appearance in the journal's opening pages, along with stories by Wyndham Lewis, but in the main the journal's material, Schmidt argues, suggested that Ford was falling back on his old friends rather than encouraging new talent. This seems a little unfair, for it overemphasises the difference of *The English Review*. In conception, *The English Review* was intended as a meeting place, and it was not, as Peter Brooker (2007) intimates, modernist *per se*; rather, it was a hybrid, a cohabitation of young and old, Edwardian and Georgian, impressionist and modern, national and international wherein change and continuity existed side-by-side. As various scholars assert, modernism did not just emerge ready-made around 1910; rather, it evolved and developed over time and place and through the work of all sorts of writers, many of whom we would now class as outside modernism (Ardis 2002; Hapgood 2000; Paxton 2000). So, while Bloomsbury would later become known as the home of modernism, in the first decade of the twentieth century the defining feature of *The English Review*, of 84 Holland Park Avenue and, more broadly, of this corner of west Kensington, was of vibrancy, cosmopolitanism and open-mindedness. It was here that a very international and progressive literary culture and aesthetic were being nurtured, and contrary to modernist-inspired conceptions of Galsworthy, it was a world in which he had a firm presence as contributor, visitor and friend.

The political atmospheres of Kensington

Progressiveness was not only the preserve of Kensington's world of letters at this time. Out on the streets and in the sitting-rooms of the neighbourhood a radical form of suffrage politics was brewing. It had its origins in the Kensington Debating

Society, a discussion group for women's rights formed in the area in the 1860s. A successor to this was the National Union of Women's Suffrage Societies (NUWSS) that by the early twentieth century had two branches within Kensington while the Women's Social and Political Union (WSPU) had both branch representation and its first shop at 143 Church Street (Sharp 1909). From 1905 until the outbreak of the First World War the activities of pro- and anti-suffragists created a heady, taut and at times quite menacing atmosphere within the area. Atmospheres, as Ben Anderson observes (2009), are the shared grounds, both literal and figurative, palpable and ethereal, through and from which subjectivities, feelings and emotions emerge. They are made as bodies mingle, affect and envelop one another in some collective, albeit often momentary, world. Thus, what went on in the streets of Kensington left few untouched. To paraphrase Kathleen Stewart (2007), it caught people up in something: enthusiasm, derision, passion, fear, comradeship and purpose. Some wanted to be caught, they wanted to be part of things, whereas others were more inadvertent captives, coming upon events, being swept up in them and having their power of action affected by the happenings of the place.

The street performances of the suffrage movement took many forms. Pageants, rallies, demonstrations, leafleting, door-stepping, and fund raising. One very visible presence within Kensington, jettisoned into the public sphere by *Votes for Women*, the newspaper of the WSPU, was the female newsy (DiCenzo 2003). Street newspaper vendors were traditionally men or boys, often from the lower social classes. The female newsy, in contrast, was a volunteer, most likely a woman of leisure who had the time to give to the cause. These women were not just sellers but 'walking and talking advertisements, demonstrating their commitment and engaging in debates with people in the street' (DiCenzo 2000: 123). They were also, very profoundly, out of place. Respectable women were domestic creatures whose innate habitat was the private sphere. Those who ventured outside it, and engaged in activities that very literally placed them in the gutter, were seen as renouncing their feminine virtue and status as gentle women (DiCenzo 2003). As one of these 'walking and talking advertisements', Violet Hunt expresses something of the tense atmosphere on the streets of Kensington at this time:

> Miss May Sinclair and myself were drafted off to hold collecting-boxes for three whole days in High Street Station . . . The arcade that led up to the station was not, at that time, a gay panorama of ladies hats, but a bare expanse on either side like the walls of a tank and perhaps that aided the terrible impression that was ours when, by concerted arrangement, May and I flashed our boxes out. Much has been said of our heroism in 'standing outside to beg,' and I fancy she felt as I did – as if we had suddenly been stripped naked, with a cross-sensation of being drowned in a tank and gasping for breath.
>
> (Hunt 1926: 41–2)

Places are often guided by tacitly known and understood rules of use, engagement and comportment, what Marc Augé (2002) might term, a contractual consensus. When this consensus is transgressed, when in Kensington of the early twentieth century women acted out of place – taking to the streets, being bold and assertive – they invited similar social contraventions from their adversaries. There were many instances of physical and emotional abuse of women like Hunt, which underscore her sense of exposure and vulnerability and bring home the contingent nature of the street's atmosphere; at once empowering and offensive. Yet there are moments of kindness. As Hunt proceeds, she and May had asked their friends to 'come and cheer us up as we stood there pilloried, and they backed us up splendidly. Mr John Galsworthy sauntered along and tipped us immeasurably, and Mr Laurence Housman and Mr Ford Madox Hueffer' (1926: 42).

Hunt's recollections point to the intersecting of trajectories on the streets of Kensington. At first sight it appears to be an intersecting that was constructive and supportive and it was; Galsworthy and Housman were two of the founding members of the Men's League for Women's Suffrage (MLWS) while Hueffer spoke and wrote in support of various suffrage groups[3]. This moment, though, also points to some of the permutations, fractures or versions of the suffrage movement that were being performed in Kensington. If we look closely at this interaction it looks somewhat passive; the walking towards, the giving of money, the possible taking of a flyer or paper and the walking on. In asking their friends for support, Hunt and Sinclair demanded some kind of response from these men, but so too did the nature of the place in which this response was enacted. Kensington High Street was a bustling place at this time with its flows of commuters, most of whom were men, shoppers, most of whom were women or servants, and the kilter of daily street life with its pattern of deliveries, its horse drawn trams and its ways through to elsewhere. Threatening for Hunt and Sinclair it was also a place of exposure for Galsworthy, Housman and Hueffer and the combination of friendship and publicity happening in this place affected what Cameron Duff (2010) terms, their power of action.

Duff (2010) argues that every encounter shapes an individual's affective capacities, either enhancing or diminishing their power to act. Each encounter, though, is always more than the meeting of two bodies for each encounter *takes place*, and, as Jon Anderson et al., (2010) argue, place has a voice that, even if faint, makes itself heard in these encounters. Thus, one's power to act depends not only on one's co-presence with others, but with the agency of place. Bearing this in mind, let us return to Kensington High Street for here we encounter one version of Galsworthy the author. Galsworthy, as we know, was a supporter of women's suffrage writing a series of articles in support of the movement in *The Nation*. Alongside this, he donated autographed books to sell in the WSPU shops and attended, along with Ada, suffrage meetings (Dupré 1976). His support though, was rational rather than radical. The series of articles Galsworthy wrote for *The Nation* offered a highly reasoned, legalistic argument for female suffrage, which was, as Lady Constance Lytton (1910) wrote in *Votes for Women*, indicative of Galsworthy's theoretical rather than active support for female suffrage. Galsworthy was not alone in this kind of support. The Men's League for Women's Suffrage (MLWS), born in the spring of 1907, had among its numbers many men of letters whose support, Angela John (1997) argues, was premised on perpetuating the chivalric, courteous and good-mannered gentleman. Aware of the weight of their words and of their public standing these men of letters, which included H.G. Wells, Gilbert Murray, Laurence Housman and John Masefield, were cautious in how and how far they expressed their support. The words of Pierre Bourdieu come to mind here, for as he observed:

> Few people depend as much as artists and intellectuals do for their self-image upon the image of others, particularly other writers and artists, have of them. . . . This is especially so for the quality of writer, artist or scientist, which is so difficult to define because it exists only in, and through, co-optation, understood as the circular relations of reciprocal recognition among peers.
>
> (Bourdieu 1993: 116)

Suffrage politics were just one cause these men lent their support to, and for many of them, and especially Galsworthy, they needed to choreograph their authorial identity very carefully so as not to alienate their audience or to compromise their other causes. They were, then, bundles of interests and when Galsworthy, Housman and Hueffer demonstrated their support for Hunt and Sinclair there was, immanent within them, other trails of action and possible trajectories (Ingold 2011).

Thus, what we need to remember about the atmosphere of place is that it is never singular nor all-encompassing; rather, there are atmospheres. These overlay one another, bleed into one another, fracture and coalesce in new ways and alter in intensity and force. They are what Anderson terms, 'spatially discharged' (2009: 80). Kensington may well have been a significant site within the suffrage movement, and one where a radical and very public politics was played out, but it also played host to other forms or trails of progressive politics. Catherine Dupré (1976), for example, recounts a dinner the Galsworthys hosted at 14 Addison Road in the autumn of 1907 to counter the censorship of Harley Granville Barker's play *Waste*. At the dinner Gilbert Murray was enlisted to obtain the assistance of J.M. Barrie and together the three writers secured an audience with the Home Secretary, Herbert Gladstone. Although little of substance came of the meeting if we return to the place of this movement's instigation, to 14 Addison Road, Holland Park, west Kensington, it is, after Ingold, a meshing of lines; a momentary coming together of trails or trajectories of actions. These trails proceed from somewhere, bringing with them the hint or scent of other places and they are becoming, moving forward somewhere else, taking something of their meeting place with them on their onward journey. Galsworthy exemplifies this so well, for at this time he was engaged not only in the cause of anti-censorship and women's suffrage but also with: sweated industries; minimum wage; labour unrest; labour exchanges; ponies in mines; divorce law reform; prison reform; aeroplanes in war; docking of horses tails; slaughterhouse reform; caging of wild birds; slum clearance; zoos; children on the stage; performing animals and the vivisection of dogs. Thus, in this particular corner of west Kensington we find all manner of trails of action and progressive politics accruing, which suggest a world irreducible to one particular atmosphere or way of being and doing. Instead, it is a world of complexity and synchronic happenings in which identity has to be negotiated.

The providential space of Holland Park

The providential space or productive locale of Addison Road was in many senses then, unremarkable. It was a space of lived social relationship where social classes comingled, local and global entwined and all manner of cultural and aesthetic politics were performed. Yet it was a space that was producing new ways of being and doing and in closing this chapter let us explore how this locale became an active voice in Galsworthy's literary making (Anderson et al. 2010). This does not mean tracing influence from page to place, or exploring how place gets folded into fiction, but rather, how novels articulate a geography that is peculiar to the time and space of their creation (Crang 2008), or, as Andrew Thacker (2003) terms it, how social space actively shapes literary form. Central to Thacker's discussion of this relationship in the early twentieth century is the contrast between two forms of literary knowledge: the map and the tour. The influence of Michel de Certeau is clear here, and is used to contrast the realist tendency to offer a vision of place from above, in which place is anchored and known through its spatial relations, with modernist texts that actualise and live space, describing it not through locations but through movements and actions.

Thacker sees the tour as a style manifest in modernist fiction in this period. Galsworthy has never been seen as a 'modern', partly because evaluations of his work rarely escape the influence of Virginia Woolf's critique in *Mr. Bennett and Mrs. Brown* (1959), but also because he tends to be judged by the whole not the parts of his literary career. Yet if we look to Galsworthy's work at this time it, like the work that was appearing in *The English Review*, was notable for its difference. During 1908 he was at work on the *Shadows*, which would later become *Fraternity*. More will be said about *Fraternity* in Chapter 8, here though let us focus on its difference, for it is a novel that has been considered something of an anomaly within the Galsworthy oeuvre. Its anomalous status arises, Dupré (1976) argues, from its modernist and experimental leanings. Unlike Galsworthy's earlier novels, such as *The Man of Property* and *The Country House*, *Fraternity* was written from a more detached point of view and focused much more on character than on story (Mottram 1956; Garnett 1934). Both earlier novels begin in houses, in the former we enter the home of old Jolyon Forsyte and spy on his 'at-home', in the latter we dine with Horace Pendyce and his guests. In each, the interior geography of these places, the layout, those present and their relations to one another are clearly laid out for the reader. *Fraternity* has no such beginning; rather, it swoops down from the sky above Kensington High Street and enters the stream of humanity that finds itself there: a cartographic flourish is quickly humanised through the characters we get caught up with in this stream. This opening sets the tone of *Fraternity*; this is not a novel confined to the domestic sphere, nor is it structured through clear relationships, significant events or known boundaries, instead, it is a novel of the quotidian life of a few inhabitants, in a small corner of Kensington.

Fraternity tells the story of the Dallison family – Hilary Dallison, his wife Bianca, his brother Stephen and Stephen's wife Cecilia – as they come to realise that their comfortable, middle class lifestyle is predicated upon the existence and toil of their working-class shadows. From this, it can be seen as a condition of England novel, but it is one more akin to E.M. Forster's *Howard's End* than the social fiction of H.G. Wells. As Thacker has suggested, *Howard's End* was a novel ambivalent about modernity; it sought stasis but encountered only flux, and in its interspersing of spaces – local and global, public and private – it attempted to capture the mobility and interconnectedness of modern social life. *Fraternity* did something similar, as we can see if we return to the novel's reoccurring street motif. Chapter 6 opens with Hilary Dallison's foray into a social world previously seen only from the bedroom window, which 'looked out at the back over some dreary streets' (1995: 47):

> Hilary . . . entered Hound Street from is eastern end. It was a grey street of three-storied houses, all in one style of architecture. Nearly all the doors were open, and on the doorsteps babes and children were enjoying the Easter holidays. They sat in apathy . . . Nearly all were dirty; some had whole boots, some half boots, and two or three had none. In the gutters more children were at play; their shrill tongues and febrile movements gave Hilary the feeling that their caste exacted of them a profession of faith: 'To-day we live; to-morrow—if there be one—will be like today'.
>
> (Galsworthy 1995: 34)

The prose is economical; the short sentences, devoid of superfluous information, and the use of semi-colons and colons help to convey how Hilary's vision and knowledge of the street unfolds as he moves through it. These are impressions caught in the moment. There is ambivalence here though too. Note for instance, the reference to boots, the cadence of the sentence would suggest that after whole boots and half boots should come no boots, but instead, Hilary's careful eye enumerates the lack of boots with more accuracy. A passer-by would unlikely be so observant, suggesting that Galsworthy was similarly trying to find a new way to express the mobility of social life, but fell back, on occasion, to the more cartographic form of realism. Later in the novel we join Hilary as he pursues one of the family's female shadows through the streets of west Kensington, in a bid to protect her from unwanted male attention. 'Certain that Hughs was really following the girl, he had but to keep him in sight and remain unseen. . . . Taking advantage of shop-windows, omnibuses, passers-by, and other bits of cover, he prosecuted the chase up Campden Hill' (1995: 59). Hilary's enumeration of his progression in terms of the utility of the busy landscape captures the interconnectedness but also the anonymity of life on the move. Paths cross, but they can also run in parallel, never quite meeting as the pace and muddle of the street conceal and obscure one from the other. This is a world of possibility; the street narrative does not convey a sense of social order or direction, but rather, reveals the 'throwntogetherness' of lived experience (Massey 2005).

Along with the flux the street settings afford, the narrative of *Fraternity* does not unfold in a linear manner. As a story of character, events (and their sequence) are relatively peripheral and this, as Bryony Randall (2007) argues, is a narrative strategy that plunges the reader into the whirligig of the moment and demands from them an ongoing making of meaning. Randall's emphasis is on drawing out the temporality of modernist writing, but it can be argued that what this also does is plunge the reader in the midst of the everyday as a lived, relational happening. The seeds of this are evident not only in how the novel opens but in the nature of the first few chapters which are taken over to establishing the lineaments of the central characters, and it is through them that we learn of the little model, a character who is largely absent or unheard, but who is central to all that follows. In the first three chapters we come to know that she is poor, that she sat for Bianca Dallison, that the picture that resulted has received much critical acclaim and that the Dallisons' view of the little model is somewhat ambiguous. This comes to us not through any particularly momentous event, gathering or authorial intervention, but through overheard conversations and brief glimpses into each character's mental interior. Musing on his responsibility towards the young model, Hilary wondered '[h]ow far was it one's business to identify oneself with other people, especially the helpless—how far to preserve oneself intact—*integer vitae*?' (1995: 23, emphasis in original).

It is in Chapter 4 'The little model' that the narrative loops back to explain the antecedents of the picture and of Hilary and Bianca's involvement in her life, and the subsequent chapters are concerned, in one way or another, with the effects and the affects the little model has on the Dallison family. The absence of events, the looping back of the narrative to earlier times and the way in which things that might otherwise have been events, such as the little model's sitting, are presented in terms of their consequences, works to add depth to the novel's happenings and to create a sense not only of ongoing temporality (Randall 2007), but of multiple temporalities: of different lives touching but rarely coinciding. At the same time, it creates a very kaleidoscopic narrative, which rarely remains in any one place or with any one character for any length of time. The novel's kaleidoscopic nature is supplemented by the use of short chapters that run, at most, to about four pages, allowing Galsworthy to shift the scene or character focus easily. As readers we experience time and space moving on, we pass through any number of everyday spaces as we do so, and we encounter, but do not linger anywhere or with anyone for too long. The effect of this is uncertainty, for no character ever comes to 'know' another. A mobile world offers too many possibilities, too many versions of self and other, and too many fissures and gulfs to overcome.

As Michael Levenson (2011) observes, the late nineteenth and early twentieth century was marked by a sceptical shift from world to self, object to subject, and, if we follow Thacker (2003), from the cartographic to the tour. Thus, the realist desire to capture the world-as-it-is was joined by what we can broadly term a modernist desire to capture it as-it-appears; mobility and consciousness became an antidote to stasis and order. *Fraternity*, it can be argued, was very much at the cusp of these two traditions. It was an experimental novel that sought to write the world in a new way, and much of its experimentalism came from the way it registered west Kensington's lived world. *Fraternity's* shadows speak of Kensington's 'throwntogetherness'; its kaleidoscopic unfolding and formal experiments hint at the area's aesthetic inventions; and, the indecisiveness and possibilities of the novel allude to the multiple atmospheres that were found on Kensington's streets in the early twentieth century. *Fraternity* does not dissolve the social and spatial differences of Kensington into a coherent whole, instead, it solidifies and accentuates them. At the end of the novel, Hilary goes abroad, not with the little model, but alone, abandoning his wife, Bianca; his attempt to cross class boundaries has only heightened the differences within and between his own class and that of the shadows. As in the throwntogetherness of west Kensington, so in the throwntogetherness of *Fraternity's* version of this neighbourhood, there is no resolution, point of stasis or refuge, only ongoingness.

Notes

1 Douglas Goldring was sub-editor for *The English Review*.
2 J. Marjoram was the pseudonym for Ralph Mottram, friend and sometime apprentice to Galsworthy.
3 One of Hueffer's most notable lectures was 'The Women of the Novelist'. This was held on 5 February 1908 and the proceeds went to the London Society for Women's Suffrage.

References

Anderson, B. 2009. Affective atmospheres. *Emotion, Space and Society*, 2 (2), pp.77–81.

Anderson, J., Adey, P, and Bevan, P. 2010. Positioning place: polylogic approaches to research methodology. *Qualitative Research*, 10 (5), pp.589–604.

Ardis, A. 2002. *Modernism and Cultural Conflict, 1880–1922*. Cambridge: Cambridge University Press.

Augé, M. 2002. *In the Metro*. London: University of Minnesota Press. Trans. T. Conley.

Blair, S. 2004. Local modernity, global modernism: Bloomsbury and the places of the literary. *English Literary History*, 71 (3), pp.813–833.

Bourdieu, P. 1993. *The Fields of Cultural Production*. Oxford: Polity Press.

Brace, C. and Johns-Putra, A. 2010. Recovering inspiration in the spaces of creative writing. *Transactions of the Institute of British Geographers*, 35 (3), pp.399–413.

British Census. 1901. *Return for Addison Road, Kensington*. Available at: www.findmypast.co.uk [accessed 19 January 2016].

Brooker, P. 2007. *Bohemia in London: The Social Scene of Early Modernism*. Basingstoke: Palgrave Macmillan.

Cassis, Y. 1985. Bankers in English society in the late nineteenth century. *The Economic History Review*, 38 (2), pp.210–299.

Crang, M. 2008. Placing stories, performing places: spatiality in Joyce and Austen. *Anglia: Journal of English Philology*, 126 (2), pp.312–329.

Dennis, R. 2008. *Cities in Modernity: Representations and Productions of Metropolitan Space, 1840–1930*. Cambridge: Cambridge University Press.

DiCenzo, M. 2003. Gutter politics: women newsies and the suffrage press. *Women's History Review*, 12 (1), pp.15–33.
DiCenzo, M. 2000. Militant distribution: votes for women and the public sphere. *Media History*, 6 (2), pp.115–128.
Drakes, C. 1999. *Holland Park Circle: Artists and Victorian Society*. London: Yale University Press.
Duff, C. 2010. On the role and affect of practice in the production of place, *Environment and Planning D: Society and Space*, 28 (5), pp.881–895.
Dupré, C. 1976. *John Galsworthy*. London: Collins.
Florida, R. 2002. *The Rise of the Creative Class*. New York: Basic Books.
Ford, F.M. [1931] 1999. *Return to Yesterday*. Manchester: Carncanet Press.
Galsworthy, J. [1909] 1995. *Fraternity*. New York: Carroll and Graf.
Garnett, E. ed. 1934. *Letters from John Galsworthy, 1900–1932*. London: Jonathan Cape.
Gieryn, T.F. 2002. What buildings do. *Theory and Society*, 31 (1), pp.35–74.
Gindin, J. 1987. *John Galsworthy's Life and Art: An Alien's Fortress*. Ann Arbor: University of Michigan Press.
Goldring, D. 1944. *South Lodge*. London: Constable and Company.
Hapgood, L. 2000. Transforming the Victorian. In L. Hapgood and N. Paxton, eds. *Outside Modernism*. Basingstoke: Macmillan, pp.22–42.
Hunt, V. 1926. *The Flurried Years*, London: Hurst and Blackett.
Ingold, T. 2011. *Being Alive: Essays on Movement, Knowledge and Description*, London: Routledge.
Ingold, T. 2008. Bindings against boundaries: entanglements of life in an open world. *Environment and Planning A*, 40 (8), pp.1796–1810.
Jacobs, J.M. and Merriman, P. 2011. Practising architectures. *Social and Cultural Geography*, 12 (3), pp.211–222.
Jenkins, L. 2002. Geography and architecture: 11 Rue du Conservatoire and the permeability of buildings. *Space and Culture*, 5 (3), pp.222–236.
John, A. 1997. Men, manners and militancy: literary men and women's suffrage. In A. John and C. Eustance, eds. *The Men's Share? Masculinities, Male Support and Women's Suffrage in Britain, 1890–1920*. London: Routledge.
Keating, P. 1991. *The Haunted Study: A Social History of the English Novel, 1875–1914*. London: Fontana.
Lang, L. 1886. The life and work of Sir Frederick Leighton, bart. *The Art Annual*, London: Art Journal.
Levenson, M. 2011. *Modernism*. Yale: Yale University Press.
Light, A. 2007. *Mrs Woolf and the Servants*. London: Penguin.
Lucas, E.V. 1906. A *Wanderer in London*. London: Macmillan Company.
Lytton, C. 1910. A message to friends and foes. *Votes for Women*, 22 April, p.479.
Massey, D. 2005. *For Space*. Oxford: Sage.
Michie, R.C. 1998. Insiders, outsiders and the dynamics of change in the city of London since 1900. *Journal of Contemporary History*, 33 (4), pp.547–571.
Mol, A-M. and Law, J. 1994. Regions, networks and fluids: anaemia and social topology. *Social Studies of Science*, 24 (4), pp.641–71.
Morrison, M. 1996. The myth of the whole: Ford's English Review, the 'Mecure de France' and early British modernism. *English Literary History*, 63 (2), pp.513–33.
Mottram, R.H. 1956. *For Some We Loved: An Intimate Portrait of Ada and John Galsworthy*. London: Hutchinson.

Paxton, N. 2000. Eclipsed by modernism. In L. Hapgood and N.L. Paxton, eds. *Outside Modernism*. Basingstoke: Macmillan.

Peppis, P. 2000. *Literature, Politics and the English Avant-Garde: Nation and Empire, 1901–1918*. Cambridge: Cambridge University Press.

Randall, B. 2007. *Modernism, Daily Time and Everyday Life*. Cambridge: Cambridge University Press.

Sauter, R. 1967. *Galsworthy the Man: An Intimate Portrait*. London: Peter Owen.

Schmidt, M. 2013. Ford Madox Ford: the good collaborator. In L. Colombino and M. Saunders, eds. *The Edwardian Ford Madox Ford*. Amsterdam: Rodopi, pp.49–64.

Scott, A. 2006. Creative cities: conceptual issues and policy questions. *Journal of Urban Affairs*, 28 (1), pp.1–17.

Sharp, E. 1909. Painting Kensington Purple, White and Green. *Votes for Women*, 12 March, pp.422.

Sheppard, F.H.W. ed. 1973. *Survey of London, Vol.37: North Kensington*. London: London City Council.

Shotter, J. 1993. *Conversational Realities: Constructing Life through Language*. London: Sage.

Sleight, S. 2009. Reading the *British Australasian* community in London, 1884–1924'. In C. Bridge, R. Crawford, D. Dunstan, eds. *Australians in Britain: The Twentieth-Century Experience*. Melbourne: Monash University, pp.71–74.

Stewart, K. 2007. *Ordinary Affects*. Durham NC: Duke University Press.

Thacker, A. 2003. *Moving Through Modernity: Space and Geography in Modernism*. Manchester: Manchester University Press.

Thrift, N. 1996. *Spatial Formations*. London: Sage.

Todd, S. 2009. Domestic service and class relations in Britain, 1900–1950. *Past and Present*, 203 (1), pp.181–204.

Woolf, V. [1923] 1959. *The Captain's Death Bed and other Essays*. New York: Harcourt Brace Janovich.

Woollacott, A. 2001. *To Try her Fortune in London: Australian Women, Colonialism and Modernity*. Oxford: Oxford University Press.

3 Posting overseas*

Author, audience and the narration of place

Introduction

The relationship between travel and writing is far from simple. Travel is a form of cultural encounter; it is an immediate and embodied immersion in difference. Travel writing, in contrast, is a space of reflection and a form of cultural translation (Duncan and Gregory 1999). Consequently, there is a tension between the travel and its mediation, for experience can never be easily translated. There is slippage and incongruence as words and things, forms and norms escape translation and encounter resistance. The result is always the narration of an imagined geography (Said 1978); a geography that is often determined more by the place of reception than by the place of inspiration. This chapter is not, ostensibly, about travel writing, and so, beginning with the above reflections may seem a little odd, yet it proffers some useful ways of thinking about how the narration of place travels – how place in fiction is alighted upon, composed and translated across space, and how this movement is facilitated or impeded – and how the place of reception is never discrete or separate from the place of creative practice.

It is to these questions that this chapter turns, doing so through an exploration of the relationship between an author and their audience. Audience here is not conceived as the specific readers we find in Chapter 8, but rather the idea of the audience, what has been termed elsewhere the intended audience, or implied reader, who haunts any literary practice (Iser 1974). This is the writer's idea of a reader, the social or spatial group to whom they think their work will appeal and to whom they seek to shape and orientate their imagined world. The intended audience becomes particularly powerful when we consider how the narration of place travels, for any representation of the 'other' will always be constrained by the social space of its reception. If a work is to find expression and receive endorsement it must resonate with the cultural norms, values and reference points of its audience (Gadamer 1975; Livingstone 2003).

Any narration of place, therefore, is a negotiation between an author's self-identity and the expectations of their audience. An author is not able to narrate place as they alone wish; what they write must resonate with their intended audience. This chapter

*An earlier version of this chapter appeared in *Narratives of Travel and Tourism* (2012), edited by T. Rakic and J. Tivers.

explores this relationship in the context of Anthony Trollope's first, and largely overlooked novel, *The Macdermots of Ballycloran*, which was published in 1847. Trollope, particularly in light of his Barchester novels, is often seen as a quintessentially English novelist[1], *The Macdermots* though, gives us a very different narrative of place, one far less certain of its identity or that of its audience. It, like this chapter, was not ostensibly a work of travel writing, but it was a novel that drew heavily from Trollope's travels around, and residence within Ireland during the 1840s. Travel, this chapter argues, became a complicating factor in the relationship Trollope had with his audience, for the more he travelled within Ireland the more he came to lose sight of who his audience was and what they expected from him in his narration of place.

Author, audience and the challenge of travel

By Victorian standards Trollope was a prolific traveller whose geographical encounters stretched from Australasia, through Africa and the West Indies to North America. Trollope was also very astute and rarely missed an opportunity to turn these travels to his commercial advantage, popularising them in the form of travel narratives and also using them as material for his fiction. It was his early travels in Ireland that were to be so central to the creation of *The Macdermots of Ballycloran*. Yet Trollope was far from a conventional traveller, for he travelled not as a tourist, but as a Surveyor's Clerk for the General Post Office. Moreover, as a civil servant, Trollope did not just travel to and around Ireland, he resided there. Trollope arrived in Ireland in 1841 and did not return to live in England for any length of time until 1859. He became part of Ireland's Anglo-Irish population, and while Daniel Corkery (1931) condemns the writing that emerged from this population as little more than travellers' tales – stories written about Ireland, by the English for the English – Trollope's inhabitation in this place complicated his narration of it.

To inhabit place, as Ingold (2000) observes, is to construct and establish roots, both material and metaphorical, within place. Ireland was where Trollope met his wife, where he brought up his children and where he became a successful civil servant and writer, and therefore he put down deep roots, making his status somewhat ambivalent, for what was 'home' and what was 'away'; was he an insider or an outsider, a native or a foreigner, a resident or a visitor? In literary terms this ambivalence was challenging for it raised questions over the location and expectations of Trollope's audience and of his ability to chime with this audience. It is this ambivalence that makes *The Macdermots of Ballycloran* so interesting, for its narration of place does not easily conform to the conventions of Anglo-Irish writing or the expectations of an English audience at this time (Corkery 1931; Trumpener 1997). Rather, it is as though Trollope uses his travels around Ireland and his dwelling within place to offer an alternative vision of Ireland, wherein popular stereotypes of the nation are simultaneously employed and challenged as Trollope tries to balance his own self-identity with the expectations of his audience.

In the nineteenth century who was the audience and what did they expect, or, in the language of Hans Gadamer (1975), what was the horizon of expectation that prevailed at this time?[2] The nineteenth century was, in many ways, a time of audience transformation, for this was when women came to be recognised as an important constituency of readers (Flint 1993). Britain at this time was a strongly patriarchal society, and the emergence of the female reader generated significant anxieties over women's reading matter and the way in which they responded to this matter. As the fairer sex women were construed as too emotional, of too nervous a disposition and too sensitive to cope with the imaginativeness of fictional writing, particularly the sensation novel (St Clair 2004). It was seen as a corruptive influence, something that would inflate their expectations and distract women from their natural calling as wives and mothers (Flint 1993).

The unease over the woman reader was not helped by a more general questioning of the novel and its affects at this time. In the later eighteenth century concerns had been raised over the moral instruction of the novel, for while a novel could have a moral message, the young or impressionable female reader may well miss the moral overtone in preference for romantic intrigue or adventure (Flint 1993). These concerns over affect were still prevalent in the nineteenth century as evangelicals began to condemn novels and novel reading as immoral and unchristian. Novels were dangerous for they took the reader into people's private lives revealing romantic entanglements and improper and indelicate situations that were damaging to the norms of female propriety (Brantlinger 1998). This was made worse by the shift in the practice of reading from a family undertaking to a private one. The growth of novels began to make reading more of a pleasurable, rather than edifying undertaking and the consequence was that reading became more uncontrollable. When reading had been a family pursuit texts had to be suitable for a family audience, the growth of the private reader meant that it was more difficult to control what people read and how they responded to it (Flint 1993; Brantlinger 1998; Secord 2000).

Alongside gender, class was also an important constituent in the nineteenth century horizon of expectation. It was not until 1870 that education became compulsory and even then this was often insufficient to establish literacy. Among the literate working classes mechanics institutes helped make texts more readily available, but these tended to mainly offer educational texts rather than fictional literature (St Clair 2004). Moreover, as Carey (1992) argues, as the century progressed and literacy did increase writers themselves became resistant to writing for the masses and attempted to exclude them through their choice of topic, language and narration. Consequently, the majority of novelists in the nineteenth century wrote for and about what they knew, and this was ultimately middle and upper class society and lifestyle. Towards the end of the century this was being challenged, but in the early decades the working class reader or novel was not widely acknowledged (Hilliard 2006).

To complement this social geography of the nineteenth century audience it is necessary to think more specifically about this audience's location. London and Edinburgh had been the main publishing centres for much of the preceding century (St Clair 2004) with many provincial towns having their own public lending library. In the nineteenth century, Mudie's Circulating Library, owned by a consortium of London publishers, came to dominate and regulate the lending market.

Rather than overseeing the extension of lending libraries and reading consumption around the country, Mudie's developed restrictive practices that ensured the monopoly of its London publishers (St Clair 2004). The result was a library that had exerted great sway in shaping not only reader expectations, but more crucially authorial approaches. Consequently, for much of the nineteenth century it was Mudie's, as audience writ large, which many authors sought to write for. If they could receive Mudie's endorsement then an audience became almost guaranteed (Griest 1970).

The narration of place is never a geographical given or expression that is merely 'put into' the text. Instead, it is a forgery, in the sense that it is both assembled and false; it is made from ideas and experiences, but these can never be mirrored in the text, partly because they will always be representations, but more importantly in this instance, because place is the outcome of the negotiations between an author and their intended audience. In the world before the text, lived experience is always tempered by a horizon of expectation, such as that which prevailed in the nineteenth century. The narratives of place that can be discerned within *The Macdermots* give us insight into this negotiation, for we can see how Trollope's travels round and residence within Ireland distanced him from the expectations of London's reading public and led him to develop a very unconventional form of Anglo-Irish place writing. *The Macdermots* was never a successful novel, suggesting that Trollope misjudged his 'audience', yielding too much to his own intentions rather than the expectations of his audience, yet it serves to focus discussion on how being-in and moving-through place can shape the relationship between an author and their implied audience.

Trollope's Ireland: a colonial tale

Trollope arrived in Ireland in September 1841, taking up his first posting in Banagher, a village on the banks of the Shannon River in the west of the country. Three years later he moved to Clonmel further to the south. During this time he was travelling extensively within the south-west region resolving postal issues, clarifying postal administration and finance and attempting to make the postal system more effective. At the same time he was also attempting to write. He started his first novel, *The Macdermots of Ballycloran* in 1843 and worked on it throughout the following year. It was completed in 1845 and published in 1847 by Newby of London. The novel tells the story of the Macdermot family. Larry Macdermot, his son Thady and his daughter Feemy once belonged to the Catholic gentry, but over the years the family had become more and more impoverished and indebted. Their plight was worsened by the arrival of the Protestant excise officer Ussher, whose seduction of Feemy led Thady, in a moment of madness, to murder him. Thady's attempts to elude the police and his subsequent trial and execution for murder led to the eventual demise of the family and the estate.

As Richard Mullen (1990) observes, this was far from a pleasant tale written as it was on the cusp of change. Famine was taking hold of the country, and this combined with poverty, religious differences and political tension revealed an

Ireland on the brink of catastrophe. As a surveyor's clerk Trollope encountered some of the worst of this distress, for his first eight years in the country were spent making 'continual journeys through its southern, western and midland portions' (Trollope 1849: 532). These were some of the most underdeveloped and least prosperous areas of the country and were to be worst hit by the potato famine (Mokyr 1983). It was during these journeys and his residence in Banaghar and then Clonmel that Trollope was 'thrown much among Irishmen of every class. I have seen them in their comparative prosperity and their too positive adversity' (Trollope 1849: 532).

Trollope's travels allow him to know. They give him a spatial knowledge that is unmatched by many of his contemporaries, and as such, he establishes for himself, and his writing, a position of authority, 'No Englishman has, I believe, had a wider opportunity than I have had of watching the changes which have taken place in Ireland during the last ten years' (Trollope 1850: 201). It is unsurprising, therefore, that Trollope is often seen as a representative of British power, a colonial voice, within Ireland and his fiction construed as little more than an extension of the administrative will to know, colonise and commodify Ireland (Corbett 2000; Deane 1994). In this light it is, perhaps, far from odd that Trollope develops such a dark theme within *The Macdermots*, for there is a need to establish Ireland's difference and discontinuity with the known, in order to justify his own role and intervention in Irish life and affairs.

Consequently, as Deane (1994) argues, one of the most persistent themes within Anglo-Irish writing is of Ireland as an anachronistic nation, whose romantic and archaic culture is in need of modernisation. The deployment of this theme serves to emphasise Ireland's difference and underdevelopment vis-á-vis Britain, and this simultaneously sanctions and endorses English intervention in the country. As Mary Corbett (2000) observes, we only need look to the first page of *The Macdermots* for this to be reinforced. The novel opens with the depiction of the old Ballycloran estate near the village of Drumsna, in Country Leitrim:

> After proceeding a mile or so . . . I began to perceive evident signs on the part of the road of retrograding into lane-ism; the county had evidently deserted it . . . Presently the fragments of bridge presented themselves, but they too were utterly fallen away from their palmy days, and now afforded but indifferent stepping-stones . . . These . . . I luckily traversed and was rewarded by finding a broken down entrance to . . . a demesne. But ah! How impoverished, if one might judge from outward appearances . . . The usual story, I thought, of Connaught gentlemen; an extravagant landlord, reckless tenants, debt, embarrassment, despair and ruins.
>
> (Trollope 1991: 1–2)

In opening his narrative with a sweeping panorama of people and place, Trollope is deploying one of the most established traits of travel writing, the all-encompassing and authorial vision. It suggests a pervasive and penetrating knowledge; the author has perfect vision, knows the history and future of a place and has the power to reveal its darkest secrets (Hillis Miller 1969). At the same time, the encompassing panorama announces Trollope's arrival; he uses it to set the scene and this, as Mary Louise Pratt (2008) observes, is a conventional way of establishing the rules of contact. It allows Trollope to assert his control over the narrative, to identify his audience as English and reassure them that the social relationships on which the story will pivot will conform to their understanding of Ireland as 'other'.

The sweeping panorama presents to us an abandoned and archaic landscape, one that has been left to run wild. Here, Corbett (2000) argues, Trollope is drawing an implicit comparison with an English rural idyll. England is the ideal of an ordered and productive landscape and, in contrast, Ireland is left wanting. Its landscape is not just different, but is degenerate and degenerating. This is a reoccurring motif within Anglo-Irish writing and owes its origin, Foster (1989) argues, to the tensions between (often absent) landowners and their tenants. At a time of land-hunger, increasing population and the endless sub-division of farmed land, the traditions of land ownership were seen as far from conducive to the nation's rural economy. Thus, in characterising Ballycloran's decline as part of Ireland's 'usual story', Trollope intimates that its desolation is to be expected. As a result, the opening of *The Macdermots* hints not at a new tale, but rather, at one that is conventional and expected by an English readership.

Trollope's Ireland: an alternative tale

It is misleading, however, to accept without question that Trollope can only encounter Ireland through a colonial mentality, given his English origins. Narratives of place, as Derek Gregory (1995) observes, are complex and often accommodate multiple and contradictory subject positions. If we probe *The Macdermots of Ballycloran* in more subtle ways the colonial voice begins to break down and there emerges, instead, a voice that is more humble and reflective. There is, James Buzard (2005) argues, a tendency to see the majority of Victorian fiction as complicit in a process of 'othering', yet such a view closes down and over-simplifies the nuances of the novel, which is inescapably autoethnographic, not in the way that Pratt (2008) uses the term but, rather, in that it provides Trollope with a way of negotiating and thinking through his own identity[3].

Trollope's move to Ireland in 1841 was very much motivated by a desire to start afresh. In London he was unhappy and uncertain of himself:

> . . . nobody thought I was right to go . . . To become clerk to an Irish surveyor, in Connaught, with a salary of £100 a year, at twenty-six years of age! I did not think it right even myself, – except that anything was right which would take me away from the General Post and from London.
>
> (Trollope 1980: 61–2)

Trollope was despised by his superiors in London, he lived beyond his means, had few friends and was beset by feelings of inadequacy and dissatisfaction (Trollope

1980). Consequently, his relocation to Ireland was a process of self-discovery; it is telling that Owen Edwards (1991) refers to Ireland as Trollope's 'frontier'. Although this hints at a colonial mentality the intimation is that Ireland is a space where Trollope can find and reinvent himself, and while Trollope never came to consider himself Irish, he likewise never considered himself an alien or a foreigner (Trollope 1849). Trollope's travels around the country were part of his process of self-discovery:

> It was altogether a very jolly life that I led in Ireland. I was always moving about, and soon found myself to be in pecuniary circumstances which were more opulent in comparison with those of my past life. The Irish people did not murder me . . . I soon found them to be good-humoured, clever – the working classes very much more intelligent than those of England – economical and hospitable.
>
> (Trollope 1980: 65)

This offers a somewhat different vision of Ireland to that supposedly prevalent within England at this time (Curtis 1997), but it reflects, perhaps, the origins of Trollope's success. Unlike his contemporaries, who found their success in English contexts and spaces and for whom any depiction of Ireland was very much a depiction by an outsider, Ireland 'made Trollope' (Corbett 2000; Edwards 1983, 1991). Ireland was a country Trollope came to know in intimate geographic detail, and, as already noted, it was where he met his wife, brought up his children, and made himself indispensible to the postal administration (Trollope 1980) and, most importantly, where he succeeded as an author. Thus, while Trollope's English audience was important to the nature of his narrative, so too was his positionality within Irish society.

Returning to *The Macdermots*, a sense of this is evident in Trollope's decision to focus on the fortunes of a Catholic family. At the time Trollope was writing, Catholicism was seen in a very perfidious light by the English. Fears of popery and the power of Rome had been sufficient to curtail Catholic rights throughout the eighteenth century, while in the nineteenth century Irish Catholicism was being conflated with nationalist elements antithetical to the Union (Foster 2001). Yet, Ireland was a predominantly Catholic country and Protestant society was limited outside of Dublin, more so because of the preponderance of absentee Protestant landowners (Hill 1997; Foster 2001). Consequently, Trollope was part of a minority religious grouping in a society that was, as he quickly discovered, deeply divided along religious lines, for 'Soon after I reached Banaghar in 1841, I dined one evening with a Roman Catholic. I was informed the next day by a Protestant gentleman who had been very hospitable to me that I must choose my party' (Trollope 1980: 72).

Trollope was to disregard this advice and developed close friendships with men like the Mayor of Clonmel, Charles Bianconi, an ardent nationalist and Catholic (Glendenning 1993). Moreover, his travels around Ireland as a surveyor's clerk moved him out of the relative security of a middle-class urban lifestyle and brought him into contact with the extreme poverty that was swelling within the ranks of the Irish Catholic peasantry (Trollope 1980). Thus, where Anglo-Irish writing often depicted Catholicism in a negative light, Trollope came to display a more sensitive knowledge and understanding of Catholicism than many of his Protestant contemporaries (Edwards 1991). This emerges quite clearly in *The Macdermots*, where Father John McGrath, the local Catholic priest and confidant to Thady Macdermot, becomes one of the novel's heroes. Father McGrath had:

> been at college at St Omer, and afterwards at Paris, and had officiated as a curé there; he had consequently seen more of French manners and society than usually falls to the lot of Irish theological students . . . He possessed also very considerable talents, and much more than ordinary acquirements, great natural bonhomie, and perpetual good temper. He was a thorough French scholar, and had read the better portion of their modern literature.
>
> (Trollope 1991: 26)

Since the founding of St Patrick's College in Maynooth in 1795 few Catholic priests had received their training or travelled outside of Ireland (Foster 2001). Thus, far from reinforcing the popular caricature of the Irish priesthood as cloistered, insular and ignorant (de Nie 2004; Curtis 1997) Trollope began to complicate it. Trollope apparently despised racial stereotypes in fiction, believing them to reduce and mask the richness of social life (Escott 1913; Tracey 1982). Furthermore, he refused to slight that which was different. As he wrote of Father McGrath:

> I have drawn as thoroughly good and fine a man as I know how to depict . . . I have lived much with the clergy men of your church, and have endeavoured to draw them in their colours as I saw them. But, because they were priests of a church which was not my church, I have never drawn one as bad, or hypocritical or unfaithful.
>
> (Trollope 1983a: 645–646)

Trollope's residency in, and travels around, Ireland, afforded him social knowledges that moved him beyond simplistic stereotypes. Instead, he was able to offer the reader a vision of the Catholic clergy as an educated, enlightened and rational influence.

On its own, this does not suggest too great a disjuncture between Trollope and an English audience, but confusion begins to arise as the relationship between McGrath and Thady Macdermot develops. It is McGrath who Thady Macdermot turns himself in to, and it is through McGrath that Thady receives some absolution for his crime. Although angry 'with the young man . . . at his folly in attempting to fly from justice' (1991: 281) McGrath regards the murder not as a premeditated act, but as an impulsive defence of his sister and his nation from English violence and dishonour. While it is one thing to challenge Irish stereotypes it is something else to depict them as disparaging of the English. It suggests that Trollope is not just narrating, but getting inside the Irish consciousness (Edwards 1991). He is losing some of his outsider status, and it is his travels around Ireland, Tracey (1982) believes, that are responsible for the development of an insider's deep and detailed knowledge of, and communion with, place. At the same time, Trollope's travels distance and detach him from England, making him, perhaps, more dispassionate about her relationship to Ireland. The result is uncertainty about whether Trollope is writing to smooth his way into Irish society or, alternatively, to educate an English audience.

This sense of ambivalence becomes even more pronounced if attention is paid to Trollope's metaphorical, as well as his spatial, travels. Trollope was an assiduous Irish scholar, and on his arrival in Ireland he renewed his acquaintance with his old schoolfellow, Sir William Gregory of Coole Park in County Galway. The house was not far from Trollope's first posting in Banaghar and it was here that he met Irish writers like Charles Lever and began, Escott (1913) claims, to study Ireland's literary canon. This canon was developing at a time when an Irish readership and a nationalist literature were growing in significance (Trumpener, 1997), and this was complicating the orientation of the nation's narration, for whom (and for where) were Anglo-Irish writers now writing: for England or for Ireland? While Tom Dunne (1988) believes that England remained the reference point for much Anglo-Irish writing up until at least mid-century, there was within the Anglo-Irish tradition a growing hesitancy and uncertainty over how, and for whom, to articulate Ireland.

Any encounter we have with a narrative of place will always frame our own experience of that place (Tavares and Brosseau 2006), and in Trollope's case this led him to depict Ireland through the conventions and tropes of Ireland's native authors (Tracey 1982). The overriding convention, or tonality, was one of simmering tension and, as Tracey (1982) rightly acknowledges, Trollope's Irish fiction, in sharp contrast to his English fiction, is brutal and violent. In *The Macdermots* alone there is murder, execution, torture, fighting and treasonable activities. This tonality was articulated through the trope of landownership, which captured the tensions between the landlord and the tenant class. Without wishing to reiterate Tracey's (1982) arguments, Trollope used this relationship not to legitimise the existing pattern of landownership and exonerate Anglo-Irish landlords, but to critique it.

Irish landownership was characterised by a network of landowners who rented out their land to tenants. In the early nineteenth century, though, this was complicated by a rural populace that was too large to be accommodated on Ireland's available land. The result was land-hunger, instability of tenure and a process of rack-renting whereby the demand for land enabled landlords to push up rents and sub-divide land to maximise their incomes (Hoppen 1998; Trollope 1850). In a series of letters to *The Examiner* on the causes of the Irish Famine, Trollope was extremely critical of this system of landownership, believing it to have been causal to the famine, and it was this critique he first developed in *The Macdermots*. In this novel Lord Birmingham is:

> a kind, a good man, a most charitable man! Look at his name on all the lists of gifts for unfortunates . . . 'Tis true he lives in England, was rarely in his life

> in Ireland . . . Could he be blamed for this? Could he live in two countries at once? or would the world have been benefited had he left the Parliament and the Cabinet, to whitewash Irish cabins . . . This would have been his own excuse . . . Yet shall no one be blamed for the misery which belonged to him?
> (Trollope 1991: 82–83)

In replicating Irish literary conventions and critiquing the system of land tenure instigated by the English, Trollope was participating in Ireland's production of its own narrative of space. The land and the soil, Deane (1994) argues, were tropes through which the Irish produced Ireland for the Irish. It was a theme that enabled them to articulate Ireland in a culturally specific, rather than a culturally generic, or English, manner. This was not, however, the only way in which Trollope moved across the frontier, as Edwards (1991) termed it, and entered the Irish consciousness. In contextualising Thady's murder of Ussher, Trollope presents Thady as a victim of political circumstances. As an excise officer Ussher was hunting 'Ribbonmen', who were accused of distilling illegal liquor. At this time the Ribbon Society was an agrarian group established to resist exploitation by the landowning classes and Thady's brief association with this group was enough for his murder of Ussher to be seen as a political act, rather than the act of defence that it was.

Trollope's intimation was that Ireland was herself a victim of circumstances. Ireland, with its agrarian violence and social tensions was viewed with suspicion and distrust by the English, and in this context any act of violence, particularly in rural areas, was seen as violence against the Union. Consequently, Trollope presented Thady as having little chance of a fair trial, for Ireland would always be judged by outsiders, not insiders. In recognising and narrating both this injustice and that of the land, Trollope displayed his affinity with Irish politics and his debt to Irish literature. His metaphorical travels through Ireland had allowed him to jettison caricatures of the nation and its people, and narrate them instead in a more nuanced and sensitive light.

Since the appearance of Sir Walter Scott's *Waverley* in 1814, novels with Celtic themes were proving popular among Victorian readers (Trumpener 1997), yet on the basis of *The Macdermots* it is doubtful that Trollope would agree. While the novel did not go unnoticed among English reviewers, they shared a belief that Trollope had lost sight of his horizon of expectation. As *Shilling Magazine* commented, 'We wish that the author had thought proper to modify his narrative, in some parts, for the roughness does not add one title to the full development of the story' (Anon. 1847: 566). Meanwhile, *The Athenaeum* advised Trollope to turn away from an Irish context for it was synonymous with terror, tragedy and misery which was unpalatable and uninviting to many readers (Anon. 1847: 517). This was something Trollope's publisher was to reiterate, noting 'that readers do not like novels on Irish subjects so well as on others. Thus you will perceive, it is impossible for me to give any encouragement to you to proceed with novel writing' (Trollope 1983: 17–18).

In seeking a London publisher and in recognising that in Ireland 'no one knew that I had written a novel' (Trollope 1980: 75), Trollope was writing not for the

Irish, but for the English. Yet, in emulating Irish literary tropes, Trollope had created a novel that was far from agreeable, for it offered a somewhat unconventional, unromantic, but in Trollope's view, authentic vision of England's Celtic periphery (Trollope 1980: 80). It was a novel in which he abandoned any sense of happy-ending idealism, experimenting instead with the anti-hero, the flawed personality and a very dark and gritty realism. The result was a work of fiction that Michael Sadleir, one of Trollope's earliest biographers, regarded as a political tract before it was a novel and an Irish novel before it was an English one. Ireland, Sadleir argued, 'having by friendliness, sport and open air saved Trollope from himself, all but choked the very genius that she had vitalised by her insane absorption in her own wrongs and thwarted hopes' (Sadleir 1927: 142–3).

It was, of course, pleasantness and happiness that English readers wanted and expected (Flint 2001), and their absence within *The Macdermots* was troubling. It was as though Trollope had poured into the novel too much of himself and his belief that Ireland was continually misrepresented by the British (Trollope 1849). Trollope had yet to learn in his fiction how to moderate his sense of self-discovery and its distillation. It was something he came to realise in later years, advising George Eliot not to 'fire too much over the heads of your readers. You have to write to tens of thousands, and not to single thousands' (Trollope 1983a: 670). Yet, in 1847 it was the message and not its narration that motivated Trollope, 'As to the plot itself, I do not know that I ever made one so good, or, at any rate, one so susceptible of pathos. I am aware that I broke down in telling, not having yet studied the art' (Trollope 1980: 70–71).

Conclusions

The Macdermots is not a piece of travel writing in the conventional sense, for it positions itself, before all else, as a novel. Yet, it is a novel that cannot escape the fact of Trollope's travels or his engagement with the narratives of place. While Trollope does not substitute his own travels as the plot, the people and places he encounters during his journeys around Ireland haunt the novel's plot, character and narration. The novel's focus on an impoverished Catholic family whose life is spiralling out of control, the predatory English in the form of Ussher, the sensible and rationale influence of Father McGrath and the overlaying of all this with a tonality of despair and tension, point towards Trollope's growing familiarity with Irish cultural and political life.

It is not, however, merely the form of *The Macdermots* that bears the indelible imprint of Trollope's travels. Trollope's journeys shaped his own sense of self; they were, in many respects, voyages of discovery. He crossed frontiers and in so doing began to re-assess, re-mould and reinvent his own identity. Although he never relinquished his 'Englishness', his travels blunted it, for while they brought him into contact with difference, his residency in place divested this difference of its abstractness and romanticism. As a result, *The Macdermots* looked not upon the difference of Ireland, but upon the *causes* of this difference, and thus the novel was problematic for its English readers as it offered a narrative of place that was far from expected. Novels, particularly those produced in cultures with firm moral codes, can rarely afford to offer a narrative of place that is unexpected, for to be unexpected implies a misjudging of the audience. The tale an author tells relies upon an audience to bring it into being, to adopt it, to support it, to read it. Yet, this audience is far from given; rather, authors rely upon response-inviting structures, which work within the novel to ensure it resonates with a particular place and its people. This, of course, is not an infallible outcome, for the implied reader is a creature of the author, emerging from authorial experience and understanding like the novel itself.

Ireland was Trollope's refuge, a place where he could remake and prove himself. It was a place for which he developed a great sympathy, since like him it had been mistreated by the English. Yet, in trying to convey the intensity of place and experience, *The Macdermots* emerged as an impassioned and impetuous response to Ireland. The novel attempted to narrate place in all its 'authenticity'; it sought to move beyond the superficialities and conventions of Ireland as narrative, to offer a form of knowledge that, in its intimacy, its 'realness' and its 'accuracy', would transform English perceptions of Ireland. In challenging conventions, however, Trollope created a narrative whose locality and particularity hindered its translation across space. *The Macdermots* indulged rather than articulated self-experience; it opted for a narrative that was introspective rather than extrospective, specific rather than general. Its difference was too extreme, with the result that it alienated rather than enlightened its intended English audience.

Notes

1 Nathanial Hawthorne once observed that Trollope's novels were as English as beefsteak, and this association between Trollope and Englishness has been perpetuated by his contemporaries such as Henry James, and in critical analyses like that of Simon Gikandi (1996).

2 James Secord (2004) has cautioned against broad generalisations of audiences, arguing that such studies forget regional and temporal specificities. However, in the world before the text it is worth remembering that the implied reader, as an imagined entity, could only ever be a broad social and spatial construct.

3 Mary-Louise Pratt uses the term autoethnographic to refer to the ways in which people describe and account for themselves through the representations that others have constructed of them.

References

Anon. 1847. New Books. *Douglas Jerrold's Shilling Magazine*, 5, pp.564–6.

Anon. 1847. New Novels. *The Athenaeum*, 15 May, pp.517.

Brantlinger, P. 1998. *The Reading Lesson: The Threat of Mass Literary in Nineteenth Century British Fiction*. Bloomington: Indiana University Press.

Buzard, J. 2005. *Disorientating Fiction: The Autoethnographic Work of Nineteenth-century British Novels.* Woodstock: Princeton University Press.

Carey, J. 1992. *The Intellectuals and the Masses: Pride and Prejudice among the Literary Intelligensia, 1880–1939*. London: Faber and Faber.

Corbett, M.J. 2000.. *Allegories of Union in Irish and English Writing*. Cambridge: Cambridge University Press.

Corkery, D. 1931. *Synge and the Anglo-Irish Literature: A Study*. London: Longman.

Curtis, L.P. 1997. *Apes and Angels: The Irishman in Victorian Caricature*. Revised edn. London: Smithsonian Institute Press.

de Nie, W. 2004. *The Eternal Paddy: Irish Identity and the British Press, 1798–1882*. London: University of Wisconsin Press.

Deane, S. 1994. The production of cultural space in Irish writing. *Boundary 2*, 21 (3), pp.117–144.

Duncan, J. and Gregory, D. 1999. Introduction. In J. Duncan and D. Gregory, eds. *Writes of Passage: Reading Travel Writing*. London: Routledge, pp.1–13.

Dunne, T. 1988. Haunted by history: Irish romantic writing 1800–50. In R. Porter and M. Teich, eds. *Romanticism in National Context*. Cambridge: Cambridge University Press, pp.68–91.

Edwards, O.D. 1991. Introduction. In A. Trollope. *The Macdermots of Ballycloran*. Avon: Folio Society, pp.vii–xlviii.

Edwards, O.D. 1983. Anthony Trollope, the Irish writer. *Nineteenth-Century Fiction*, 38 (1), pp.1–42.

Escott, T.H.S. 1913. *Anthony Trollope: His Public Services, Private Friends, and Literary Originals*. London: John Lane.

Flint, K. 2001. The Victorian novel and its readers. In D. David, ed. *The Victorian Novel*. Cambridge: Cambridge University Press, pp.17–36.

Flint, K. 1993. *The Woman Reader, 1837–1914*. Oxford: Oxford University Press.

Foster, R.F. 2001. *The Irish Story: Telling Tales and Making it up in Ireland*. London: Penguin.

Foster, R.F. 1989. *The Oxford History of Ireland*. Oxford: Oxford University Press.

Gadamer, H.-G. 1975. *Truth and Method*. London: Sheed and Ward.

Girkandi, S. 1996. *Maps of Englishness: Writing Identity in the Culture of Colonialism*. New York: Columbia University Press.

Glendenning, V. 1993. *Trollope*. London: Pimlico.

Gregory, D. 1995. Between the book and the lamp: imaginative geographies of Egypt, 1849–50, *Transactions of the Institute of British Geographers*, 20 (1), pp.29–57.

Griest, G.L. 1970. *Mudie's Circulating Library and the Victorian Novel*. Bloomington: Indiana University Press.

Hill, J. 1997. *From Patriots to Unionists: Dublin Civic Politics and Irish Protestant Patriotism, 1660–1840*. Oxford: Oxford University Press.

Hilliard, C. 2006. *To Exercise our Talents: The Democratisation of Writing in Britain*. London: Harvard University Press.

Hillis Miller, J. 1969. *The Form of Victorian Fiction*, London: University of Notre Dame Press.

Hoppen, K.T. 1998. *The Mid-Victorian Generation, 1846–1886*. Oxford: Oxford University Press.

Ingold, T. 2000. *The Perception of the Environment: Essays on Livelihood, Dwelling and Skill*. London: Routledge.

Iser, W. 1974. *The Implied Reade; Patterns of Communication in Prose Fiction from Bunyan to Beckett*. Baltimore: John Hopkins University Press.

Livingstone, D.N. 2003. *Putting Science in its Place*. London: University of Chicago Press.

Mokyr, J. 1983. *Why Ireland Starved: A Quantitative and Analytical History of the Irish Economy, 1800–1850*. London: Allen and Unwin.

Mullen, R. 1990. *Anthony Trollope: A Victorian in his World.* London: Duckworth.
Pratt, M.L. 2008. *Imperial Eyes: Travel Writing and Transculturation.* 2nd edn. London: Routledge.
Sadleir, M. 1927. *Trollope: A Commentary.* London: Constable.
Said, E.W. 1978. *Orientalism*. London: Penguin.
Secord, J. 2004. Knowledge in transit. *Isis*, 95 (4), pp.654–672.
Secord, J. 2000. *Victorian Sensation: The Extraordinary Publication, Reception and Secret Authorship of Vestiges of the Natural History of Creation.* London: University of Chicago Press.
St Clair, W. 2004. *The Reading Nation in the Romantic Period.* Cambridge: Cambridge University Press.
Tavares, D. and Brosseau, M. 2006. The Representation of Mongolia in Contemporary Travel Writing: Imaginative Geographies of a Travellers' Frontier. *Social and Cultural Geography*, 7 (2), pp.299–317.
Tracey, R. 1982: 'The unnatural ruin': Trollope and nineteenth-century Irish fiction, *Nineteenth Century Fiction*, 37 (3), pp.358–382.
Trollope, A. 1983. *The Letters of Anthony Trollope*, Vol.1, *1835–1870.* N. John Hall, ed. Stanford: Stanford University Press.
Trollope, A. 1983a. *The Letters of Anthony Trollope*, Vol.2, *1871–1882.* N. John Hall, ed. Stanford: Stanford University Press.
Trollope, A. [1883] 1980. *An Autobiography.* M. Sadleir and F. Page, eds. Oxford: Oxford University Press.
Trollope, A. 1850. The real state of Ireland. *The Examiner* 30 March 1850, pp.201–2.
Trollope, A. 1849. Irish distress. *The Examiner*, 25 August pp.532–533.
Trollope, A. [1847] 1991. *The Macdermots of Ballycloran.* London: Folio Society.
Trumpener, K. (1997). *Bardic Nationalism: The Romantic Novel and the British Empire.* Princeton: Princeton University Press.

Part II

Writingscapes

Writers at work

4 Bennett's writingscape

On Friday 27 October 1899, a day on which the London newspapers were full of reports on the second Boer War that had begun a little over a fortnight earlier, Arnold Bennett boarded a train in London bound for Torquay. He was escaping the city for a long weekend sojourn with his friend the dramatist-writer, Eden Phillpotts. This weekend, as shown elsewhere (Saunders 2008), was something of a watershed in Bennett's practice, who recorded how his creativity started to develop when in the companionship of Phillpotts (27 October 1899, 1932: 96). The earlier work explored the sociality of this weekend and how being within a social world was instrumental to the evolution of Bennett's writing practice. This chapter returns to this weekend to understand more fully its significance to the development of Bennett's writing routine or what can be termed his 'writingscape'.

Writingscape is used to encapsulate the practical nature of writing of which the putting of pen to paper is only one aspect. It is a term that gestures to Tim Ingold's (1993) concept of the taskscape and to a lesser extent, albeit a more directly literary one, Howard Gruber's (1989) notion of the network of enterprise. Beginning with Gruber's work, a network of enterprise is an individual's writing strategy: the creative agenda, the proposed lines of activity, the daily pattern of work and the temporal organisation and recurrent practices that allow a particular task or activity to happen and be completed. These are the mundane happenings of writing, the routine and unexceptional things that are rarely commented upon but which give shape and substance to a writer's life. The taskscape is similarly concerned with the array of practical operations, acts and activities that constitute and comprise our being-in-the-world, but it is also more nuanced in that it probes the temporality of the taskscape. Time within the taskscape is social, meaning that at work within the world we are always attending to one another. This is important, for where some have conceived the taskscape as unthinking repetition (Edensor 2003), the social time of the taskscape means 'we continually feel each other's presence in the social environment, at every moment adjusting our movements in response to this ongoing perceptual monitoring' (Ingold 1993: 65). Doing, therefore, is always a coupling of social perception and movement, and this means that tasks never happen in precisely the same way twice, or as Harold Garfinkel puts it, they always happen for 'another first time' (1967: 9).

In Bennett's writingscape there are many couplings of perception and movement. The most obvious would be the putting of pen to paper, where each word is related and shaped by those that went before. It is, though, to a more recognisable form of movement, yet one not immediately associated with writing, that we turn here; namely, walking. On the first visit to the weekend of 27–31 October 1899 mention was made of the several walks that Phillpotts and Bennett took together and the way these brought together and shared physical worlds and mental worlds in creative stimulation. During these rambles, Bennett was particularly struck by Phillpotts' detailed botanical knowledge and his deep engagement with his local world yet he was also fascinated by the place of walking within Phillpotts' daily routine, or writingscape. As we shall see, walking was not absent from Bennett's writingscape at this time, but it occupied a qualitatively different position; a position that suggests something about Bennett's writing experience and about how he sought to develop his writing skills.

'This was really all he did': Friday 27 October 1899 to Tuesday 31 October 1899

Eden Phillpotts was a well-known author and highly voracious writer. From the early 1890s he had produced, on average, one work of fiction per year, alongside a wealth of plays, poetry and articles. In 1899 Bennett's career was not yet on a par with this; he had one published novel (*A Man from the North*, 1898), had one in process (*Anna Tellwright* later to be published as *Anna of the Five Towns,* 1902), and was juggling fiction writing with freelance journalism and the editorship of *Woman*, a weekly magazine. As a result, Bennett's practice of fiction writing at this time was a little haphazard and opportunistic. Not only was his journal peppered with the difficulties of writing – the absence of the desire to write (29 September 1896), the 'weeks of indolence' (28 January 1897, 1932: 30), and the periods of unproductiveness (11 and 23 January 1898) – but writing had to fit round other happenings, and so, was often begun in the twilight hours (29 September 1896). What Bennett had little of, therefore, was time to write. His writingscape was strongly determined by the affordances of his social world (Ingold 1993); the responsibilities of work, the demands of family and the expectations of friends meant that Bennett had to make opportunities for writing, squeezing it in and cutting it short according to the rhythms of his life.

This might explain the sense of bemusement in the first journal entry Bennett made during his stay:

> Talking of work, Phillpotts said that he worked at his serious stuff from 10.30 or 11 to 1, and sometimes in the afternoon again for an hour or so, after going for a walk. This was really all he did. Every month he interrupted the big work, to write a short story, which would take 2 or 3 or 4 days. He showed me his diary of working quantities. Up to date this year his total was 360,000 words. He had hoped to write 600,000.
>
> (27 October 1899, 1932: 96)

There are two aspects of this that repay closer attention. The first is Bennett's choice of phrase, 'This was really all he did', for it suggests a certain awe, on Bennett's part, over the ease and effortlessness with which Phillpotts wrote. It also points to a gulf in practice between the pair, a gulf which increases as the diary entry progresses and Bennett registers both the exacting nature of Phillpotts' methodological approach and the prolific output it could generate. In the earlier visit to this weekend emphasis was placed on the conversation that underpinned this diary entry and the manner in which this dialogic encounter gave Bennett a new insight into writing. It was an insight that began to dispel writing's practice as something exclusive and situated, and replace it with an awareness of its ordinariness and its spatiality. There is little doubt that Bennett admired Phillpotts' way of working, writing a few days after his return to London: 'I have now decided to act on Phillpotts's advice, to write a short story every month. I finished my November story this month: "The Phantom Sneeze", a humorous ghost tale, 4,500 words' (5 November 1899, 1932: 98). What is more, as 1899 drew to a close Bennett, emulating Phillpotts' practice, recorded his literary output for that year:

> I have written 335,340 words, grand total. 228 articles and stories (including 4 instalments of a serial of 3,000–7,500 words each) have actually been published . . . I have written six or eight short stories not yet published or sold . . . Also the whole draft (80,000 words) of my Staffordshire novel 'Anna Tellwright'. My total earnings were £592 3*s*. 1*d*
>
> (31 December 1899, 1932: 101)

The second aspect was the walk that came just prior to 'That was really all he did'. Not only did Phillpotts' routine demonstrate that writing could be conducted in a measured and methodical manner, it also intimated that inhabiting the world was as important as inhabiting the study. As Rebecca Solnit (2001) observes, the relationship between walking and writing stretches back a long way; for William Wordsworth walking was as much a mode of composition as it was a mode of travel, for John Keats it was a means to refine one's artistic sensibilities and for John Clare it proffered a narrative structure. Ingold (2011), meanwhile, presses for walking as a form of circumambulatory knowing; we know as we go and the nature of our knowing is shaped by the nature and place of our footwork. When we walk at our natural speed, in an unhurried and relaxed manner we are most likely to generate a rhythm of thinking (Solnit 2001). Furthermore, such footwork in familiar locations lends itself to contemplation, for the known world demands less attention; here one is enfolded within the landscape, it is like a second skin, known, felt, familiar and secure, freeing one's mind to wander elsewhere. This is mind-walking not world-walking and to return to Ingold (2011) this kind of walking is creative; it brings forth new worlds (although as we shall see it does not always seal the self within a hermetic, mental world). The fact that Phillpotts seemingly walked at approximately the same time and for a similar duration each day suggests that he returned to the same routes with relative frequency and moved at a steady pace. It is likely that his familiarity with these routes meant that walking was not a search for inspiration, but rather, a means of thinking; a way to abstract oneself from the moment and turn inwards. In 1899 this kind of creative walking appears quite novel, perhaps even a little indulgent, to Bennett, for whom writing was a more pressurised and harried undertaking.

'On the spot geography': November 1896 – September 1897

Although Bennett may well have been a little confounded by the way walking afforded Phillpotts a creative thinking space, Bennett's own practice at this time was not wholly confined to the study. It too was a practice that was in-the-world, but to understand this worldliness we need to temporarily fast-forward to 1914 and the publication of Bennett's guide to creative writing, *The Author's Craft*, which appeared that year. By this time, Bennett was a well-established writer and the guide opened with the following: 'A young dog, inexperienced, sadly lacking in even primary education, ambles and frisks along the footpath of Fulham Road, near the mysterious gates of a Marist convent' (1914: 3). In little over a page the dog is dead, run over by a motor-bus, and the rest of this short chapter explores the effects of this event upon the passers-by. Bennett uses this small, and to some quite inconsequential occurrence, which begins and ends in an instant, to signal the centrality of active observation to writing. A happening like this has a back-story and a future story and in the first section of the guide, 'Seeing Life', Bennett underscores the importance of being-in-the-world to effective writing, for a writer 'ought to embrace all phenomena in his [sic] curiosity' (1914: 19). In encouraging the writer to embrace the wholeness of existence, there is a disciplined way to grasp this and it begins with 'geographical knowledge'. This Bennett proceeds to argue 'is the mother of discernment, for the varying physical characteristics of the earth are the sole direct terrestrial influence determining the evolution of original vital energy' (1914: 21).

Brian Hudson (1982) has discussed at length Bennett's geographical interests and his belief in the importance of geographical study and knowledge to creative writing. By the time Bennett came to write *The Author's Craft* the nature of this study was twofold. On the one hand, it was an academic pursuit of information from a variety of sources. When living at Trinity Farm he called on his friend John Rickard for geological data on the region, desiring to 'get hold of that (in compressed tabloid form), because until I do, I can't grasp the scenery' (9 October 1900, Bennett 1968: 138). On another occasion he wrote to Hugh Walpole enclosing a copy of a conspectus of place and time, what we might otherwise know as a gazetteer, recounting how invaluable an earlier edition had been to his own compositions (Bennett 1913). On the other hand, it was a practical undertaking that depended upon, 'realistic, verifiable, on the-spot geography. I believe in always beginning close at hand – it is a trick that gives actuality, and actuality excites extraordinary interest' (1918: 213). This often translated, particularly in Bennett's early writing career, into fact-finding missions (18 April 1899).

Bennett's journal entry for 4 November 1896 gives an early account of one of these missions. At this time, Bennett was contemplating a second novel and recorded walking 'the streets of the West End, for three hours, 8.30 to 11.30, from Tottenham Court Road to Hyde Park Corner. . . . I collected nothing of any interest whatever. For over a week I had noticed nothing, made not a single useful observation' (1932: 22). In a meditation on knowledge-making, Ingold (2011) likens knowing to two modalities of movement: transport and wayfaring. In the former we are passengers moved across the world in pre-determined ways and on pre-established routes. What happens on the way and the landscapes we move through are of little importance, what matters is the points of departure and destination. It is goal-orientated. The latter, in contrast, is a responsive and reactive engagement with the world, where the path of movement is determined in the moment of its happening. Here process is more important than outcome. In some respects, Bennett's walk in London was evocative of the former. There were spatial and temporal parameters. It was purposeful albeit not in the sense of reaching a physical destination but rather, in disclosing or accessing what Bennett would later term, the 'formula' of place (24 October 1897, 1932: 54) and it could be measured and evaluated. Yet, Bennett's account of his rambles also suggests quite a responsive way of walking. It was unlikely that Bennett would have moved either head-down or rapidly around the West End; rather, it was more probable that he was looking about him and being directed by interesting leads. Add to this the distance from Tottenham Court Road to Hyde Park Corner, which was only 1.5 miles. At a moderate pace this was a journey time of about 30 minutes, but Bennett was walking for three hours, suggesting that he meandered around the area with little by way of a route plan.

This kind of walking was very much in the world not in the mind; the world provided the raw materials from which a story could be hewn. Another example of the way Bennett enlisted walking into his writingscape came less than a year later when, on 2 September 1897, he received the news that his sister Tertia's fiancée, Willie Boulton, had died in a swimming accident. In France at this time, Bennett returned to the family's home in the Potteries the following day. Although a relatively regular visitor to his family, he had been living away from the region for nine years. This visit however, turned out to be a relatively long one and while preparations were made for the funeral Bennett occupied himself, and sometimes his sister, by taking walks in and around the area. Some of these he recounted in his journal:

> when I have been taking early morning walks with Tertia [his sister] and when I have been traversing the district after dark, the grim and original beauty of certain aspects of the Potteries, to which I have referred in the introduction to 'Anna Tellwright', has fully revealed itself for the first time. Before breakfast, on the heights of Sneyd Green, where the air blows as fresh and pure (seemingly) as at the seaside, one gets glimpses of Burslem and of the lands between Burslem and Norton, which have the very strangest charm.
>
> (10 September 1897, 1932: 46–7)

The modality of these walks is suggested in Bennett's choice of phrase: traversing. Traverse implies zigzagging, criss-crossing and meandering. There is no set destination or route, but rather a desire to let place speak (Anderson et al. 2010). As such, a pathway may be picked up only to be abandoned as something catches Bennett's eye, seducing him in a new direction. Encountering new viewpoints may slow or direct movement, while turns and returns offer not a new perspective as such, but what James Gibson terms, a 'path of observation' (1979: 195); a mobile inhabitation of the world. Another meaning of traverse is to examine closely, to spend time and expend energy in place so that, as Bennett observes, things reveal themselves. This is perhaps moving closer to the kind of walking Phillpotts undertook, but unlike Phillpotts Bennett does not give himself up to the landscape. There is a purpose to the walk, there is something to be gained or extracted from it. Often this something was a discrete and identifiable narrative entity; for instance, as Bennett mentioned, his early morning walks appear in *Anna of the Five Towns* (begun as Anna Tellwright) as follows:

> Five contiguous towns – Turnhill, Bursley, Hanbridge, Knype, and Longshawe – united by a single winding thoroughfare some eight miles in length, have inundated the valley like a succession of great lakes . . . They are mean and forbidding of aspect – sombre, hard-featured, uncouth; and the vaporous poison of their ovens and chimneys has soiled and shrivelled the surrounding country . . . nothing more seemingly remote from romance. Yet be it said that romance is even here – the romance which, for those who have an eye to perceive it, ever dwells amid the seats of industrial manufacture.
>
> ([1902] 2001: 24–25)

In *A Man from the North* (1898), the opening chapters recount what was likely one of Bennett's first walks through London, while *Clayhanger* begins with Edwin standing on the 'steep-sloping, red-bricked canal bridge, in the valley between Bursley and its suburb Hillport . . . To the east rose pitheads, chimneys, and kilns . . . To the west, Hillport fields' ([1910] 2000: 31). Geography comes first, both in the real world and the imagined world and walking, both for Bennett and his characters, is how this geography is apprehended. Walking is a way of generating narrative space, and as such, it is not a regular or daily undertaking, but something that occurs in response to the demands of the narrative; it comes before not in the midst of writing.

'Lessening the risk of distraction': May 1907 to 1912

If we fast-forward to 1907 and return once more to Les Néfliers, we find Bennett engaged in a different type of walking. Frederick Marriott and his wife Margaret were some of the first friends Bennett made on moving to London, and, when Bennett moved to France the couple were regular visitors. In recollecting the current of Bennett's life at Les Néfliers, Marriott noted that:

> [Bennett] rose regularly at 6.30, and after a light French breakfast . . . went for a walk in the forest to think out the subject matter for his days [sic] writing.

> In order to lessen the risk of distractions, he took the same route each day. We know [sic] the direction and ascent of his perambulations, and when we took our morning walk, we were careful to avoid interference with his seclusion.
> (Marriott no date)

By this time, Bennett had turned to writing full time. He had given up the editorship of *Woman* in 1900 and, with more time at his disposal, gradually began to adapt his way of doing things. His journal continued to record his 'on the spot geography' but alongside this, there was a sense of the generative potential that arises from being-in-the-world. Bennett's entry for 12 May 1904 was indicative of this observing that, 'I walked a good deal about Paris yesterday, arranging instalment 4 of "Hugo"'. Similarly, in early 1906 he recorded that after 'having written my T.P.'s W. article to-day I went out for a stroll through Paris. . . . I have had several days of regular unhurried work lately, interspersed with such strolls' (29 January 1906, 1932: 226). These, along with Marriott's recollection, are just some examples of an emerging pattern of behaviour, a new routine if you will, that was becoming evident in Bennett's journals from the early years of the twentieth century. None of these was an instance of collecting or finding material for his novel, but rather, of thinking-in-motion.

As already suggested, thinking-in-motion suggests a different kind of engagement with the world. It is an engagement with the world not as a source of inspiration but as a launch-pad to the world of the imagination: it facilitates and makes possible an inward turn. Consequently, it is often associated with a form of walking that is intentionally repetitive, both spatially and temporally, conducted in this way so that, as Tony Williams suggests, the landscape is experienced as a 'blank space', an 'invisible space', in which one can mull-over, think-through and consider absent things rather than present things (2013: 234). Thus, unlike his walks in the Potteries where, like the practices of such luminaries as Wordsworth he took from the world, using walking as material and inspiration, at Les Néfliers Bennett brought to the world. If we follow Williams, what Bennett brought was an imagined world, enfolding this into, so that it almost obscured, the material world of his route. Interestingly though, and unlike Williams, Ingold (2011) does not see thinking-in-motion as occurring within a blank space. It may well be creative, allowing one to access and bring forth new worlds, but it is not a mode of walking that is wholly oblivious to time and place or productive of a closed epistemological system, for there is always, Ingold (2011) argues, difference *within* repetition: a seamless performance is only achieved through continual correction and adaptation to a world on the move. The body is always registering the world, and its awareness and responsiveness to this world increases rather than decreases with experience. The world therefore is not a blank space but an active other or influence within the writingscape: it affects how and what we do.

Its blankness is filled by the constant motion of things: by weather, by people, by the passing of time, and, if we follow a Spinozian notion of affect, being-in-the-world brings us into relational encounters with all of these things, shaping our power of action (Duff 2010) and making the world in the moment of encounter. Let us pursue these encounters further, beginning with the weather, which Ingold (2011) observes is not something we perceive, but something that envelopes us. As Robert Macfarlane (2013) remarks of the poet and nature writer, Edward Thomas (1878–1917), weather, and particularly the rain, is something he thinks in; it cloaks him, masks him almost, hiding him from other people and things. Rain is a hermetic seal, uniting self and world in deep communion. Bennett similarly enjoyed walking in the rain, recording that on 'Thursday it rained all day and I walked two hours in the rain' (5 October 1907, 1932: 259). It was a 'landscape of soaked leaves and thick clouds and rain – nothing else. But I like it' (9 October 1907, 1932: 261). Later in the same month he records 'Nothing but rain. I walked 4 miles in 59 minutes this morning in the rain. And this afternoon I went with Marguerite to Moret in pouring rain. A promenade on a thoroughly bad day in autumn is the next best thing to a promenade on a fine late spring morning. I enjoyed it immensely. I enjoyed splashing waterproof boots into deep puddles' (16 October 1907, 1932: 262). Although this afternoon walk was an instance of walking-with, it is revealing that Bennett registers the weather and the pleasure he takes in it, rather than his company. If we follow Phillip Vannini et al. (2012), weather is constantly making and remaking place. Rain transforms hard surfaces into boggy ones, shady spots into dark corners and trickles into torrents. More than this though, as Vannini et al., go on to observe, we learn to weather or dwell with these meteorological changes, adapting our behaviour, our dress and our mind-set so that we can master the environment. Bennett's boyish enthusiasm for splashing through puddles, his donning of waterproof boots and the way he embraces the rain, rather than being pushed inside by it, demonstrate how the weather shapes our taskscape: it 'occasions activities, provides us with resources, facilitates certain types of actions and discourages others, and prompts us to learn certain kinds of techniques for everyday living' (Vannini et al. 2012: 9). Walking in the rain is qualitatively different; we are aware of the elements on our skin and our clothes, it alters the pressure of our tread, our direction and our sense of surety. We are more aware of the moment, the immediate, the close-at-hand, suggesting perhaps that our route inwards is more difficult as meteorological happenings – that misplaced step that allows a trickle of water to breach the top of a waterproof boot – jolt us back into the living moment.

It was not just the weather that shaped Bennett's writingscape, so too did the other people implicit within it. Take the manner in which Bennett's daily walk depended upon its interweaving or, as Ted Schatzki (2011) terms it, its harmonisation with the timespaces of others. As Marriott noted, the other inhabitants of Les Néfliers, Marriott, his wife and Bennett's wife, Marguerite, organised their own spatialities and temporalities to avoid impinging upon Bennett's routine. Thus, the materialisation of the world as a blank or invisible space was rarely the creation of a sole individual. It depended upon the comportment of others, and at this stage in Bennett's career his family and friends were willing to adapt and suppress their timespaces to his in order to help nurture his craft. This was not always to be the case. The Bennetts' marriage was turbulent from the start. Bennett would frequently leave Marguerite to her own cares while he pursued his interests. During the first three months of 1910 Bennett's journal records the couple's stay in Brighton where Bennett penned much of *Clayhanger*. What comes across clearly is Bennett's absorption in his task. We hear little of Marguerite, beyond the odd afternoon tea or walk, and instead encounter days, like January 4, when Bennett wrote letters until 11am, went out pottering, wrote until 1pm and then continued with his writing after lunch and dinner. In the early years of marriage Marguerite accepted this, but gradually became more intolerant of Bennett's neglect and his expectation that she and others should bend to his routine (Bennett 1917). What we see, therefore, is an unravelling of the social relations that had been necessary to the development of a routine in the first place. From 1917 onwards Bennett registers more and more frequently the manner in which Marguerite's behaviour or a little more euphemistically 'household difficulties' (10 April 1918, 1932b: 226), were disturbing his writing and prompting him to develop different modes of doing and inhabitation.

In a final example of how Bennett's writingscape was shaped by external forces let us turn to the seasons. Bennett's usual walking route took him through the forest of Fontainebleau which is a mixed deciduous forest of approximately 25,000 hectares in which oak, Scots pine and European beech swath a landscape of gorges and granite rock formations (www.fontainebleau-tourisme.com). Unlike its evergreen counterpart, this is a forest that 'presents us with natural history: a vast scene of sprouting, budding, leafing out, flowering, fruiting, passing away, passing life on' (Rolston 1998: 166). Seasonal change can be felt, seen and experienced and for a man who wandered in this forest nearly every day Bennett would have felt the changes underfoot, smelt them in the air, seen them in the flora and fauna, and sensed them on his skin. As the springy grasses of summer gave way to the waterlogged paths of winter, as spring colours turned into autumn ones and woodland glades became chilly nooks, this was a place of continual transformation. Our awareness of seasonality and of place, Tim Ingold and Tehri Kurtilla (2000) observe, inheres in this relationship between rhythms of growth and the movement of animals, plants and social life; we know and feel how change happens across the year and we always tune our movements to the movements of this living world. Thus, as Ingold (2011) argues elsewhere, we always make our way through a nascent world not a preformed one, and as the world sprouts and changes around us we are unable to move through it as a hermetically sealed entity. Instead, it demands that we adjust ourselves to its shifting pressures, terrains and resistances, and in this way, our movement is always a form of participation in which perception and action are coupled (Abram 1996; Ingold 2011).

Thus, Bennett's daily walk in the forest, despite following the same route, was never only a mind walk for he could never completely turn his back on the living world. Autumn, as he noted in his journal, was his favourite season (28 October 1907) when all was 'yellow and brown. Leaves falling continuously. Horse-chestnuts quite yellow. Sound of water occasionally dislodged from the trees by wind' (23 October 1907, 1932: 263). It was during the summer though, that we see very clearly how perception and action combine:

> On Saturday morning I walked for two hours in the forest, and completely finished the construction of the Five Towns play. I saw a big, strong deer that crashed fiercely through thick bushes like a stag out of Sir Walter Scott, stood watching me some time, and then crashed and crackled off again. Once I sat down to rest on a felled three, and a squirrel appeared. It bounded to within a few yards of me then sprang up a tree, hung to the bark like a fly for several seconds, and finally ran higher very rapidly, at the same time taking to the side opposite to me so that I could not see it. The tree was not a large one nor had it either many branches or much foliage, but I could not find the squirrel, though I walked round and round it and searched most carefully.
>
> (22 July 1907, 1932: 237–238)

Although engaged in thinking through his next work, Bennett was far from immune to the living world around him: it caused him to pause, to sit, to observe and to move in response to its rhythms. Thus, Bennett was not moving through the forest as an impermeable entity, physically present yet consciously absent; rather, his body was arrested and possessed by the forest, swept up by its currents and happenings. Seasonal change though can make even 'the most well-known places . . . "flip over" and turn strange and hostile, leaving the traveller lost and bewildered' (Ingold and Kurtilla 2000: 189). The forest is where this bewilderment can be most pronounced, for the known can suddenly look and feel strange depending upon time of day or time of year. Bennett wrote of Fontainebleau's 'persistent fogs lasting till afternoon' (26 October 1907, 1932: 264) and when he first moved to the area he noted the forest's autumnal gloom, of getting lost and of his need to 'steer by the sun' (23 November 1904, 1932: 198). Bennett's disorientation is not on a par with Gaston Bachelard's (1969) evocation of the forest as a deep, bewildering and limitless world, but it did require 'a receptivity to the specific shapes and textures of . . . [the] present and a spontaneous creativity in adjusting oneself. . . . to those contours' (Abram 1996: 50). Thus, no matter how well known and how well trodden Bennett's routes through the forest of Fontainebleau were, this was a world on the move and, as such, it demanded conscious attunement and adaptation on Bennett's part.

Being on the move in a mobile world brings in to tension the sense of routine as something enduring, repetitive and unthinking. As we have seen, Bennett may well have followed the same route each day, but different things were happening on this route, altering it and demanding conscious engagement from Bennett the walker. Thus, although Frederick Marriot's recollection implies unthinking engagement, suggesting that Bennett's immediate world was, after Gilles Deleuze and Félix Guattari (2013), a striated one – a homogenous place, where things were ordered and bounded, subject to clock time and known through Euclidean geometry – there is, folded beneath this, a smoother space. This has no pre-given layout but is a patchwork of variation, a space of movement and flux in which roaming not route-following is the condition of being. Smooth spaces are ones of affect, things happen here and while these happenings may appear similar day-in-day-out, closer attention reveals all sorts of intersecting materialities that may shake, move, disturb or arrest us. These intersecting materialities have implications on production, for they are never stable, meaning that 'any series of processes are not exactly repeatable, but only approachable' (Woodward et al. 2010: 276). Thus, while the space of Bennett's daily walk may have appeared familiar and his conduct through it routine, if we look again and more closely, all sorts of variances emerge that make each moment different to the last. It was different not only because the thoughts and ideas that Bennett transported with him were different each day, but because the way these intersected with the intensities of the world was never predictable or stable. It is impossible to know how Bennett's real and imagined worlds intersected during these walks, but we can suggest that Bennett's writingscape was an active outcome of his on-going and emergent encounters with the world.

The space-time of the writingscape

The taskscape, and by implication the writingscape, is inherently social, improvisational and flowing; one is swept up, joins with and responds to the forces and flows at work in the world. Within it we find professional and personal acquaintances, seemingly stable forms of practice and modes of temporal organisation. As we look more closely though, it is an unpredictable and ongoing space: our social others shift and alter; our actions even if repeated as in a routine or habit, are never repeated in the same way twice, for they unfold within a social world and their doing must resonate with the rhythms and happenings of this world; and the place of our doing is not an enduring or stable backdrop to our action, it is always complicit within our doings. Thus, how one performs a task, in this case writing, depends upon the particular presences, processes and practices happening and ongoing at any one time.

This is not to suggest that a task happens in starkly different ways on a daily basis, rather, it should draw our attention to the nuances of change, and in concluding this chapter there are three subtleties in Bennett's writingscape that repay closer attention. The first is what we might term the temporalities of skill. This chapter opened in the autumn of 1899 and looked back on the higgledy-piggledy nature of Bennett's writingscape in which his creative labours were discontinuous, irregular and required clear environmental cues. As the twentieth century took root though, a steadier working rhythm, greater evidence of opportunity for intense concentration and a subtler engagement with the landscape emerged. Ingold (2011) argues that these are indicators of the shift from novice to expert, but more than this, they point to how the writingscape is an unfolding mental landscape if you will. It is always creeping in new directions, gradually abandoning some ways of doing, becoming more efficient at others thereby enabling time to be clawed back, and accumulating new ones that are conducive to, or generative of the time-space of the moment. Secondly, and related to this, much of this expansion and contraction owes to Bennett's shifting possession of time. When Bennett was able to turn to writing full time he came into possession of time, and this, it can be argued was productive of new spaces and modes of creativity. It gave Bennett the opportunity to give himself up to, rather than merely take from the world. If, in his novice years, he saw Phillpotts' quite measured creative practice as bemusing, the pruning of his own writingscape enabled him to adapt his ways of doing things, engaging with the world in a co-productive manner rather than merely as source material. We see this in the changing significance of walking within the writingscape as it moves from something *a priori* to, or early within the creative process, to something intrinsic to its daily momentum. Finally, taskscapes can imply a relatively constrained geography: the artist in their studio, the carpenter in their workshop or the ironmonger in their foundry, or it can refer to actions that occur within the same locale or social world. The daily commute or the walk to school, for instance, will follow the same route through place day-in-day-out. Walking, though, allows us to see how Bennett's writingscape was more expansive and supple. It too functioned through movement, through Bennett's meanders through place, but the location and locution of these meanders was, as the Potteries and the West End attest, varied. What is more, some of these meanders might not have happened every day. The fact-finding missions while not one-offs happened at particular moments within the process of writing. They tended to come near to the start of a new writing project or as the predecessor to particular developments of plot. Thus, and running somewhat contrary to conceptualisations of the taskscape, happenings within the writingscape were not always regular, spatially or temporally. In fact, it was a space that simultaneously brought together the routine and unusual, the familiar and strange, the known and unknown, folding these together in ways that could have unpredictable effects on Bennett's creative labours.

Walking is only one aspect of the writingscape, but it is one that brings us to Bennett's creative labours from a slightly oblique direction, approaching from the social world not the imagined one. As argued elsewhere though, one of the most efficacious ways of exploring practices that elude or are closed to us is to look at how practices nest within and illuminate one another (Saunders 2010). Creativity is an imaginative undertaking, and so often it is only the physical outcome, the end product, that we can access, but walking gives us insight into the lived experiences that affect, spur, motivate and impede these labours. Of course, this suggests that tasks have no clear boundaries and always bleed into one another, but it is the nature of this bleeding that is productive of the writingscape's inimitableness over time and space. In leaving Bennett's writingscape it is perhaps easiest to comprehend it as almost akin to a silicone mould; it has a shape or outline, but this is not rigid and unyielding, instead it is supple and pliable, able to respond to the different ingredients, mixtures, and consistencies of the things that come together within it. It is a productive space, in which one may anticipate the form, but not its nature, content or dexterity.

References

Abram, D. 1996. *The Spell of the Sensuous: Perception and Language in a More-Than-Human World.* London: Vintage Books.

Anderson, J., Adey, P and Bevan, P. 2010. Positioning place: polylogic approaches to research methodology. *Qualitative Research*, 10 (5), pp.589–604.

Bachelard, G. 1969. *The Poetics of Space*. Boston: Beacon.

Bennett, A. [1910] 2000. *Clayhanger*. London: Penguin.

Bennett, A. [1902] 2001. *Anna of the Five Towns*. London: Penguin.

Bennett, A. 1968. *Letters of Arnold Bennett II: 1889–1915*, J. Hepburn ed., Oxford: Oxford University Press.

Bennett, A. 1932. *The Journals of Arnold Bennett, 1896–1910*, N. Flower ed., London: Cassell.

Bennett, A. 1932b. *The Journals of Arnold Bennett, 1911–1921*, N. Flower ed., London: Cassell.

Bennett, A. 1918. Try geography for example, *The Bookman*, 23, pp.213–214.

Bennett, M. 1917. *Letter from Marguerite Bennett to Arnold Bennett*. George Beardmore Collection. A-13. Keele: Keele University Library.

Bennett, A. 1914. *The Author's Craft*. London: Hodder and Stoughton.

Bennett, A. 1913. *Letter from Arnold Bennett to Hugh Walpole*. AB 1994.5. Stoke: Potteries Museum and Art Gallery.

Deleuze, G. and Guattari, F. 2013. *A Thousand Plateaus*. London: Continuum.

Duff, C. 2010. On the role of affect and practice in the production of place. *Environment and Planning D: Society and Space*, 28 (5), pp.881–895.

Edensor, T. 2003. Defamiliarizing the mundane roadscape. *Space and Culture*, 6 (2), pp.151–168.

Fontainbleau Tourisme n.d. *Official Website of Fontainbleau's Tourist Office*. Available at: http://www.fontainebleau-tourisme.com/ [accessed 19 August 2016].

Garfinkel, H. 1967. *Studies in Ethnomethodology*. New York: Prentice Hall.

Gibson, J.J. 1979. *The Ecological Approach to Visual Perception*. Boston: Houghton Mifflin.

Gruber, H. 1989. Networks of enterprise in creative scientific work. In B. Gholson, W. Shadish, R. Neimeyer and A. Houts, eds. *Psychology of Science*. Cambridge: Cambridge University Press.

Hudson, B. 1982. The geographical imagination of Arnold Bennett. *Transactions of the Institute of British Geographers*, 7 (3), pp.365–379.

Ingold, T. 2011. *Being Alive: Essays on Movement, Knowledge and Description*. Abingdon: Routledge.

Ingold, T. 1993. The temporality of the landscape. *World Archaeology*, 25 (2), pp.152–174.

Ingold, T. and Kurtilla, T. 2000. Perceiving the environment in Finnish Lapland. *Body and Society*, 6 (3–4), pp.183–196.

Macfarlane, R. 2013. *The Old Ways: A Journey on Foot*. London: Penguin.

Marriott, F. n.d. *Adventures with Arnold Bennett*. George Beardmore Collection. AB Misc 2. Keele: Keele University Library.

Rolston, H. 1998. Aesthetic experience in forests. *Journal of Aesthetics and Art Criticism*, 56 (2), pp.157–166.

Saunders, A. 2010. Literary geography: reforging the connections. *Progress in Human Geography*, 34 (4), pp.436–452.

Saunders, A. 2008. The practice of the text: arts of conversation, arts of writing. *Historical Geography*, 36, pp.94–111.

Schatzki, T. 2011. Timespace and the organisation of social life. In E. Shrove, F. Trentmann, R. Wilk, eds. *Time, Consumption and Everyday Life: Practice, Materiality and Culture*. Oxford: Berg.

Solnit, R. 2001. *Wanderlust: A History of Walking*. London: Penguin.

Vannini, P., Waskul, D., Gottschalk, S. and Ellis-Newstead, T. 2012. Making sense of the weather: dwelling and weathering on Canada's rain coast. *Space and Culture*, 15 (4), pp.361–380.

Williams, T. 2013. The writer walking the dog: creative writing practice and everyday life. *American, British and Canadian Studies Journal*, 20 (1), pp.224–238.

Woodward, K., Jones, J.P. and Marston, S. 2010. Of eagles and flies: orientations towards the site. *Area*, 42 (3), pp.271–280.

5 Trollope's work plans

Crafting *The Bertrams*

Trollope, as the introduction to this book makes clear, was a man who became, on the posthumous publication of *An Autobiography* (1883), quite infamous for his writing routine. *An Autobiography* was a controversial publication. *The Spectator* (1883) praised its frankness and defiance, its desire to be true to the self irrespective of the conventions of the form, yet noted that Trollope was perhaps a little too headlong and impetuous in this pursuit. *Macmillan's Magazine* (1883b) was more direct, observing that Trollope's candidness brutalised the literary ideal. Both reviews were making reference to the autobiography's exposé of Trollope's writing practice:

> I found it to be expedient to bind myself by certain self-imposed laws. When I have commenced a new book, I have always prepared a diary, divided into weeks, and carried on for the period which I have allowed myself for the completion of the work. In this I have entered, day by day, the number of pages I have written, so that if at any time I have slipped into idleness for a day or two, the record of that idleness has been there, staring me in the face, and demanding of me increased labour.
>
> (Trollope 1991: 119)

John Sutherland (1982) has explored one instance of this writing practice, attending to the manuscript material of *The Way We Live Now* – Trollope's writing diary, chapter summaries and character profiles –, in order to trace the novel's construction over the course of 1873. *The Way We Live Now* was published in serial form during 1874–75. It was, as Sutherland observes, Trollope's longest novel for some time, running to 380,000 words, and was almost two-thirds longer than the triple-decker form to which he was habituated. In consequence, the demands of the form, the tight timetable for composition (Trollope began composition on 1 May 1873 and had to complete the first part of the publication for October of that year) and the need for intermediary outputs (*Harry Heathcote of Gangoil* was written in June and July of 1873) were productive of a particular way of writing which the term mechanical, so frequently applied to Trollope's working, somewhat undervalues. It is not the intention to rehearse Sutherland's excellent work here, rather, this chapter begins with this work in order to underscore the highly routinised nature of Trollope's writing practice. This chapter stays with the manuscript material that documents Trollope's writing but explores the spatialities of what appears to be a very routine way of proceeding.

The main point of departure for this chapter is the growing field of materialist hermeneutics. This was a term coined by Jerome McGann (1991) to account for meaning as something that arose not merely through the reader but through book production. Within this body of work the spatial layout of the text has emerged as a particularly significant carrier of meaning and this has solicited geographical analysis, most notably in the work of Robert Mayhew (2007). Where McGann was interested in imaginative texts arguing that these called attention to their own textuality, Mayhew turns his attention to what he terms spatial texts (geography books, expeditionary and travel accounts), arguing that spatial devices matter equally to the making of 'factual' meaning. Taking inspiration from this body of work, this chapter examines the material hermeneutics of neither imaginative nor scientific texts, but rather of draft materials. It argues that if spatial layout matters to the production of meaning of printed material it also matters to the making of this material in the earliest phases of creation. The precursor to this work is Sally Bushell's (2009) study of text-as-process. In this, Bushell argues for draft materials as conditional spaces of meaning that are made through the coming together of compositional context, authorial intention and nonintentional forces. The layout of this material – the inclusions of illustrations, marginalia, deletions and insertions – is expressive of the practice of writing. Of interest here is not so much the drafts of particular novels, for Trollope was known for his refusal to revise his work, but rather for what the spatial form of his working diaries and work plans express.

The Way We Live Now (1875) was written in the latter years of Trollope's life and, as Sutherland points out, the pre-publication material that survives for it is particularly rich. If we go further back though, to some of the earlier novels Trollope wrote, like *The Bertrams* (1859), there is a strong continuity of practice in his composition of both novels. Like *The Way We Live Now* a detailed work plan exists for *The Bertrams* that indicates that Trollope began working on this novel on 1 April 1858. According to the note at the top of the work plan Trollope was in Egypt at this time (where he was engaged in negotiating a postal treaty in his capacity as a Post Office surveyor) and continued to work on his manuscript during his subsequent travels in the Mediterranean region where he visited Malta and Gibraltar (to inspect their postal services) before journeying back to Ireland via England and Scotland. The novel was finished on 17 January 1859 whilst Trollope was in the West Indies where he had been sent in November of the same year to organise the region's postal services[1]. In this sense, *The Bertrams* is Trollope's most peripatetically written novel. Trollope's geography of composition will be returned to later, but for the moment it is the form of his plan that is of interest.

The page of *The Bertrams*

Trollope's work plan for *The Bertrams*, like his subsequent work plans, was composed on lined octavo paper. The sheet was divided into two columns and each column was then sub-divided into a further four columns: one large and three small (see Figure 5.1). Essentially then, the page was divided into eight columns, but it is important to recognise that the page was organised vertically not horizontally: one read down the first four columns and then down the next four columns. The first column of the four recorded the day's date; the second the numbers of the pages written that day or indication of why pages were not written; the third the amount of pages per day and in the final column the running total of pages written. This method of accounting was then repeated on the other half, or second column on the page. Dividing the page in this way was economical in that it permitted twice the number of weeks of practice to be visualised simultaneously, while enabling comparisons between weeks and months to be made with relative ease. Mary Hamer regards the work plan as a 'goad to action' (1987: 42) or what Michel Foucault (1982) might term, a disciplining force: a way for Trollope to administer and monitor his practice. This will be returned to later, but for the moment let us pause with the form of the plan. Its tabulated and columnar form helped Trollope adhere to a very particular manner of writing. There are, according to Klaus Hurlebusch (2000), two types of creative writing: work-genetic writing and psycho-genetic writing. The first regards writing as a communicative act that anticipates an audience from the start. As such, it is externally aware, teleological and often linear in nature. It is a form of writing that tends towards purposeful reproduction, in which the work is imagined as an entity before being translated into writing. In contrast, the second type of writing is far more internalised and discontinuous. An audience is neither sought nor explicitly imagined, instead thought is externalised as a series of workings and revisions that gradually accrete into something larger and more holistic. Although it is likely that most writers fall somewhere between the two, Trollope's work plan suggests a mode of writing very much in line with work-genetic writing[2].

The plan is drawn up in advance to comprise the number of weeks Trollope believes each project will take. Nearly all the plans were drawn up on the same octavo size paper, with Trollope giving himself one leaf of paper per novel, which yielded four pages. The first three pages were each drawn up to accommodate ten weeks per page, giving Trollope, on average, 30 weeks per novel. The final page was usually given over to chapter headings, summary or character outlines. Thus, it is not the case that writing will go on until the novel is concluded, rather the novel will be concluded in the time Trollope sets aside for it. By drawing up the plan in advance rather than adding rows as he went along, Trollope pre-fabricates time as space and this, as Stuart Sherman (1996) argues, is prognostic; it calls into being possible futures, demanding a form of writing that has a forward and successive propulsion. This spatial form, Sherman contends, arose during the seventeenth and eighteenth centuries when time-telling became ever more accurate. Instead of registering time as signal events occasioned by sun or season, diarists, such as Samuel Pepys, were able to capture the fullness of time in a much more serial and continuous manner. The diary as a result became simultaneously retrospective and anticipatory. Trollope's work plan is not a diary in the Pepysian tradition, but its spatial form is similar. It is comprised of provisionally blank,

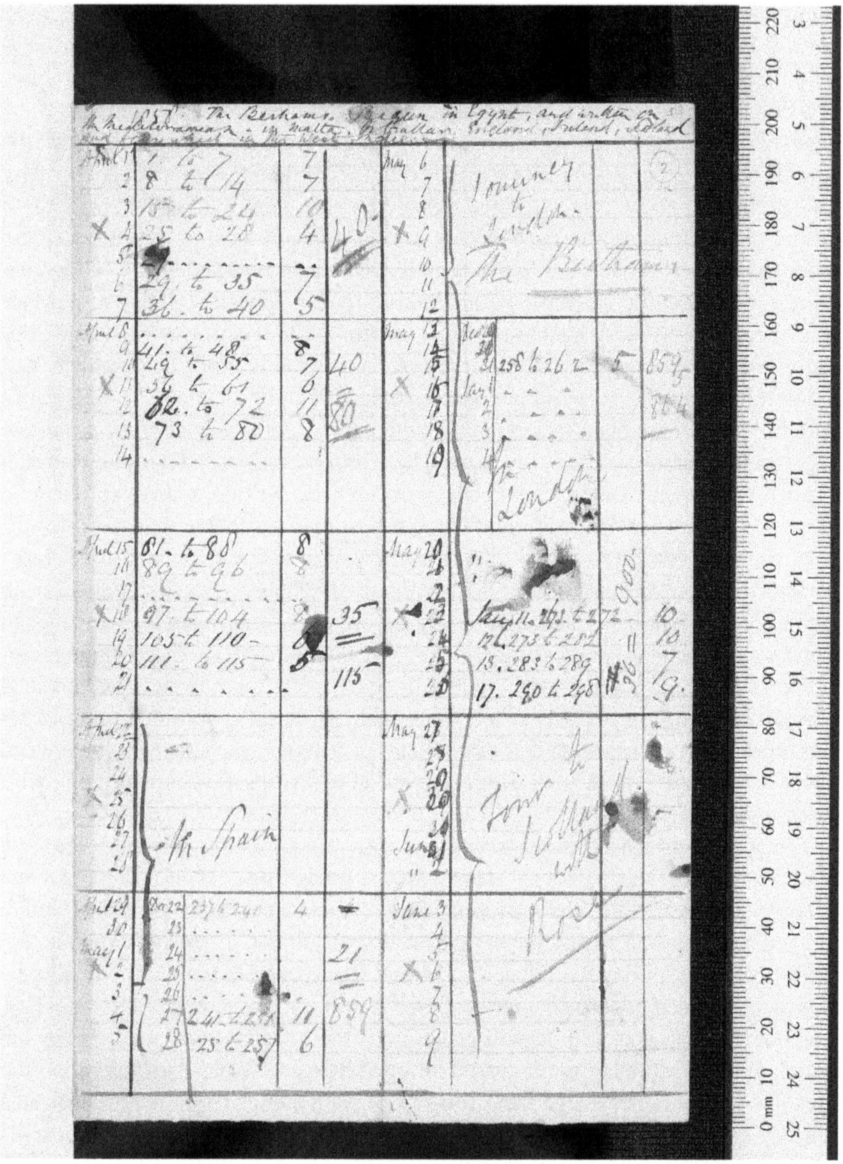

Figure 5.1 Work plan for *The Bertrams*, 1859.
Image courtesy of the Bodleian Library, MS. Don c. 9, fol.53r.

uniform and successive durations, each of which is dated and follows chronologically the one preceding it. The blank spaces of the work plan figure Trollope's time as empty and for his life to unfold these blanks both necessitate and compel daily in-filling: registering writing not as a series of occasions or moments, but as an ongoing occurrence. The cramped dimensions of these blanks are not designed to invite copious notes but a factual record of output, nor though, are they designed to permit variance from an established way of proceeding: Trollope writes from the beginning, from chapter one, he takes up each day where he left off the previous day and he stops writing only on completion of a pre-defined section. Thus Trollope's writing, like the neatly lined rows and columns of the work plan, unfolds linearly and logically over time and space. Although Trollope could pause mid-way through a number, such action is discouraged by the conventions of the form. In none of Trollope's work plans does he record partial completions, such as a page and a half, rather, he always works in whole numbers. Similarly, while Trollope could invert his consecutive numbering, following number 7 with number 12 for example, he rarely did so for this confused the logic of the form and rendered it meaningless as a measure of pages written[3].

Geographies of the page

Turning from the meta-level to the micro-level there is, running across the top of work plan, a heading space and in the top left hand corner Trollope recorded the year and the title of the work: '1858. The Bertrams.' The nature of this heading space had evolved. In Trollope's earliest work plan, drawn up in 1856 for *Barchester Towers*, there was no real heading space. The year and volume number were recorded at the top of the first week's work and the title was scrawled up the left-hand side of the page, almost as an afterthought. In *The Three Clerks* (1858), which came next, Trollope progressed to an underlined title. It is *Doctor Thorne* (1858b), which followed, that had the first distinct and separate heading space. Within this box, the title is central and around it are several numerical notes that allude to the dimensions of the undertaking: projected numbers, pages or words per volume. This was a habit that became common in the majority of Trollope's work plans, especially after he gave up writing novels in three volumes in favour of serial publication. Serial publication demanded a more careful calculation of words to ensure a relatively regular offering per serial number. Thus, the later work plans, of which *Phineas Redux* (1874) was indicative, utilised the heading space as a way of managing the material space of the printed page. This work plan begins: *80 Chapters;* 20 numbers at 48 pages = 260 words or 15 numbers at 68 pages = 260 words. In this instance, numbers refer to what we would today term issues or editions of a journal or serial. When Trollope drew up the work plan for *Phineas Redux* it was obviously undecided whether it would come out over 20 or 15 numbers, but if 20 numbers each number was to consist of 48 pages and if 15 then 68 pages. In either scenario, each page would comprise 260 words. This part of the work plan therefore, evolved to become an integral part of Trollope's conception of his book or serial as a material reality. It operated as a guide and a prompt to the scope of the writing, existing *a priori* to the writing of the novel and thereby assisting him in visualising the textual whole, dividing his labour accordingly and offering a spatial measure of productivity.

Thus, the heading space emerges as part of what we might, after Gérard Genette (1997), term the paratext of the work plan: one of the structural devices that along with the columns and boxes enable the diurnal form to appear and be recognised as such. In occupying the top of the page and running across its entire width the heading space operates in a similar way to a novel's title page: a device for naming, directing and disclosing what is to come. The information contained within this space is presented as stable and certain, and gives the appearance of being a necessary first step for writing to take place, for it makes clear an intention. In returning to *The Bertrams*, it is evident that Trollope had not yet finessed the heading space in the directive manner of his later work plans, and there was no need to, for this novel was conceived and written as a work of three volumes. It did however, like *Dr Thorne* before it, have a very distinct heading space and within this, the first words that Trollope obviously wrote were, as already noted, the date and title: '1858. The Bertrams.' Unlike in the work plans for *The Three Clerks* and *Dr Thorne* this title was not in the centre of the heading space but in the top left-hand corner. In one sense, the full stop after *The Bertrams* seems quite final, as though this was initially intended to be the only thing to appear within this space. In another, the heading space is much wider than any of the other rows within the work plan. This is partly to designate its importance, but it could also be suggested that the blank space that appeared after *The Bertrams* came to compel in-filling. Blank spaces as we know are anticipatory, but unlike the blanks spaces within the body of the work plan the blank space of the heading space was not one that demanded daily completion; rather, it was a space intended to be completed before work began as a frame for what would follow. Thus, in formalising the heading space within the work plan of *The Bertrams* and moving the date and title to the top left-hand corner, Trollope may well have been anticipating or recognising the scope this space could have in shaping the written word. He was creating, in other words, a textual space of opportunity wherein he could lay out his spatial intentions for the work that would follow.

Interestingly though, what follows '1858. The Bertrams.', is 'Begun in Egypt and written on the Mediterranean . . .', with Trollope proceeding to list all the locations of writing. Putting aside the geography for the moment, this is interesting not because Trollope makes no reference to the spatial dimensions of the work but because of his turn of phrase. Trollope writes in the passive tense, suggesting that the heading space was completed post novel or as an on-going register of the novel's peripatetic composition. Thus, far from existing prior to the content of the work plan, it is likely that the heading space of *The Bertrams* was continually being updated. This complicates its paratextual significance, for rather than codifying the spatial intentions of writing or unambiguously directing the shape of the novel, this space became another register of duration, albeit a geographical rather than temporal one. It was also a space that quickly became inadequate for such a purpose, with the places of writing pushing at the boundaries of the heading space and spilling over its material confines. It is impossible to know whether Trollope intended the space to be used in this way, but the space he allotted to the heading space does suggests that he did not intend for the novel to have such a long or peripatetic construction as it did.

The ambivalence and difference of this space vis-à-vis Trollope's other works plans may suggest that the meaning and purpose of the work plan was still not refined. This was only Trollope's fourth work plan and each was a little different in form to the last. That said, the degree of difference with the introduction of geographical information and an *a posteriori* approach to in-filling, was quite stark compared to both earlier and later plans. To the extent that, unlike most of Trollope's work plans that dealt almost exclusively in the world of the novel – in numbers, pages, plots and projections – that for *The Bertrams* was more concerned with the worldly than the imagined; it was heavily annotated with the happenings of the world outside the text. Quite frequently, for instance, the time-space of the work plan registered Trollope's location rather than pages written, sometimes hiatuses in writing were substituted with 'Ah me!' while at others the diary was suspended for weeks at a time. Thus, rather than expressing the accumulation of textual space, the work plan was expressive of the contingent and relational nature of creation. Elsewhere, Marcus Doel (1999) and David Couch (2010) might consider the time-space difference of this work plan in terms of 'spacing'; that is, as a crack in the everyday, the usual and the expected, where things are unsettled and resettled in new ways by novel energies and encounters. What this does is disrupt the linear flow and forward propulsion that the form of the work plan connotes. On first sight the work plan affirms a work-genetic mode of writing, yet the temporal and geographical framing of the work plan calls our attention to the scene of making and to the people, places and times that are gathered therein and which can confound, trouble and divert even the most resolved of routines.

The geographies of writing

In his autobiography, Trollope reflected on his process of writing and of the importance within this of imaginative construction, or what he termed, 'castle-building' ([1883] 1992: 43). It was not that Trollope spent long periods of time 'castle-building' believing such work to be mentally wearing; rather, there were usually,

> some hours of agonising doubt, almost of despair . . . or perhaps some days. And then, with nothing settled in my brain as to the final development of events, with no capability of settling anything, but with a most distinct conception of some character or characters, I have rushed at the work as a rider rushes at a fence which he does not see.
>
> (1992: 175)

This suggests a somewhat *ad hoc* and organic approach to storytelling, but it was an approach underpinned by the composition of plot outlines. The manuscript material for both *Lady Anna* (1874) and *The Way We Live Now* for instance, includes plot summaries and character outlines. It was not, however, an approach that always worked and on several occasions Trollope '. . . encountered what, in hunting language, we call a cropper. I had such a fall in . . . The Bertrams' (1992: 175). *The Bertrams* was begun on 1 April yet Trollope only finished his previous novel, *Doctor Thorne*, the day before. This may account for part of his 'fall', for there was obviously little time for imaginative construction. However, this presumes that Trollope could only hold one story in his head at any one time, which we know from his practice of working on projects simultaneously (Sutherland 1982) to be untrue. It is the work plan that discloses something more about this 'fall'.

If we look to the work plan for *The Bertrams* this fall is clearly evident in a number of ways. Trollope worked tolerably well during the first three weeks of composition, writing on average seven pages a day. By 21 April he had written 115 pages of the novel. At this point though, Trollope broke off from writing and did not return to the novel for nearly two months, picking up his pen again on 17 June. The intervening two months are not left as blank spaces within the work plan, rather Trollope records his various travels: 'In Spain'; 'Journey to London' and 'Tour of Scotland with Rose'. Nearly two months of no writing was highly unusual in Trollope's work regime. Travel could, perhaps, be construed as causal to this rupture in practice, but early on in his career Trollope had commissioned a special desk to aid his writing whilst on the move. In this sense, the condition of travel was unlikely, in itself, to have been a disruption. Moreover, the fact that Trollope continued to write something in his work plans during this period, writing on one occasion 'The Bertrams' as if to remind himself of what he was supposed to be doing, suggests that he did not lay writing aside completely and may well have tried to write something. This then, was the first fall that Trollope experienced and to understand its cause it is instructive to return to 21 April and the 115 pages Trollope had completed by this point.

Given that Trollope considered a page to comprise 260 words a rough estimate suggests that page 115 brings the reader to top of page 71 in the World's Classic edition, just several pages shy of the end of Chapter VI 'Jerusalem'. In this chapter the main protagonist, George Bertram, quits his uncle in London and arrives in Jerusalem in search of both his father, Sir Lionel, and a religious calling. The reader does not meet Sir Lionel for another two chapters, but they do encounter, amongst a number of more minor characters who will pepper this part of the story, Caroline Waddington, to whom Bertram will later become engaged. It is on page 71 that Caroline learns that Bertram is likely to inherit his uncle's large fortune. By the time Trollope reaches this part of the narrative he had introduced his principle characters and themes – the clash between the spiritual and the secular, the communal and the individual –, established his two intertwining storylines: the courtship of Arthur Wilkinson and Adele Gauntlet and that between Bertram and Caroline, and created the first narrative node of the story[4]. It could be suggested therefore, that on reaching this point Trollope had laid the foundations of his story; all it required now was building. The problem though, was how to build it and it was a problem that had evidently plagued Trollope before. Towards the end of *Barchester Towers* (1857) for instance, his intrusive narrator asked, 'who can apportion out and dovetail his incidents, dialogues, characters, and descriptive morsels, so as to fit them all exactly into 930 pages, without either compressing them unnaturally, or extending them artificially' (1994: 481). The cause of the problem was the triple-decker novel and how to manage the story within the prescribed strictures of this material space.

On average, triple-decker novels had to be around 160,000 words in length spread over 45–50 chapters, with approximately 15 chapters per volume, published in post-octavo format with a set price of 31*s* 6*d* (Gettmann 1960). A triple-decker novel, therefore, was a physical and commercial entity before it was an aesthetic one, and this conditioned a particular kind of narrative development. The idea was that the first volume should introduce the key characters, establish possible storylines and, as in the second volume too, conclude with a dramatic episode that whets the audience's appetite. The second volume was where the storylines were brought to a head before being resolved in the final volume[5]. What is more, this physical form fostered a particular kind of textual geography, one that Caroline Levine (2009) has likened to a network of relational characters and storylines. The length of the form meant that authors could explore the criss-crossing of social life, creating multiple nodes and hubs within the story to facilitate all sorts of socio-spatial interconnections and possibilities. In *The Bertrams* though, Trollope struggled to manage this network creating, in Michael Sadleir's (1945) view, a very lopsided and cumbersome novel. The work plan gives some expression to this lop-sidedness, revealing through its pauses and presences the narrative nodes Trollope felt confident in writing. Caroline's discovery of Bertram's likely inheritance is one such node, for it paves the way to their engagement, and the relative ease with which Trollope wrote up until this point suggests that the story was comparatively clear in his mind. After this, it is evident that Trollope was unsure over how to develop and build the story further.

The work that followed this, the last few pages of 'Jerusalem' and the subsequent chapter 'The Mount of Olives', which were eventually written between 17 and 21 June appear, then, to have presented Trollope with problems. There are several reasons why this might be so. The first revolves around Trollope's conception of the novel's plot. Trollope composed the book's first node relatively easily (Bertram's likely inheritance of his uncle's fortune), he similarly made light work of the second node (Henry Harcourt's visit to Littlebath) and the third node, which was the critical interview between Bertram and Caroline, wherein they broke off their engagement. This came in Chapter V 'Juno' of the second volume. After this, Trollope had his second fall and all his work plan records, for several weeks, is 'Ah me!'. Having bought the story to this point, Trollope, Mary Hamer (1987) observes, was struggling to find a new impetus for it. Thus, it seems that Trollope had a relatively clear outline of the key nodes, or turning points, within the story. Where he ran in to difficulties was with, what Franco Moretti (2006) terms, the fillers. If novels are composed of nodes and hubs then fillers are what link the decisive moments of the text together. Fillers are narration of the everyday, events of little consequence, and they proliferated in the nineteenth-century novel, Moretti argues, in an attempt to rationalise the novelistic world, reducing surprises, taming uncertainty and softening the flow of events. Furthermore, fillers helped to pad out the physical space of the three-decker novel, enabling a one volume story to be eked out into three. If we return to Trollope's first fall that arose with the latter pages of Chapter VI 'Jerusalem' and Chapter VII, 'The Mount of Olives', these chapters are interesting for they are more travelogue and religious tract than story. In these pages the reader joins Bertram as he visits Jerusalem's Church of the Holy Sepulchre and ascends the Mount of Olives. Such chapters led one reviewer to remark that 'Trollope uses his traveller's journal too lavishly' and 'inserts many things that, to say the least, are superfluities' (Anon. 1859). Trollope then, may well have conceived of *The Bertrams* as a series of nodes; scenes of heightened dramatic action upon which the storyline pivoted, but his compositional problems arose with how to link and move between these pivots. It was, then, the narration of the everyday that became vexing, leading Trollope to turn to place as both a filler and a means of moving the action onwards.

A second reason for Trollope's fall may have something to do not with fillers *per se*, but with the nature of the filler. *The Bertrams* appeared between *Doctor Thorne* and *Framley Parsonage*, two of Trollope's most enduringly successful works. *Bentley's Quarterly Review*, a short-lived competitor to the *Edinburgh Review*, considered Trollope's novels 'light literature' (Anon. 1859c: 456), and these two novels, David Skilton (1993) argues, while attractive placed few demands upon the reader. In contrast, *The Bertrams*, Skilton argues, was a much more experimental and socially resonant novel that explored contemporary intellectual debates over faith, doubt, science and materialism. It calls attention to these debates from the opening chapter which rebukes the usurping of morality, integrity and love by the competitive spirit. If Bertram's friend Arthur Wilkinson represents Christian morality and Henry Harcourt self-interest, Bertram occupies the uncertain middle ground and, as Russell Perkin (2009) observes, the two most explicitly religious chapters, wherein Bertram ponders his religious beliefs, are 'Jerusalem' and 'The Mount of Olives'. The problem with these chapters was not only that they introduced 'half serious discussions of difficult subjects' but that Trollope did not make his own views clear thereby, in the opinion of *The Saturday Review*, diluting his responsibilities as an author (Anon. 1859b: 369). That Trollope struggled with these chapters may, therefore, owe to the controversial nature of the filler and his uncertainty over how to frame debates over the liberal challenge to theology so as not to alienate his audience. The moral responsibilities of the mid-Victorian writer were high. There were, as Skilton (1972) argues, significant conventions determining moral debate within fiction and making religious orthodoxy clear was one. *The Bertrams* did not encounter such vehement criticism for its failure on this score as Trollope's later works, but the fall in the work plan coinciding as it does with religious debate, suggests that Trollope struggled with how to frame Bertram's renunciation of a clerical life without getting mired in religious controversy. If this is the case, the aporias in Trollope's work plan reveal the fold lines between worlds; the complexities of the real world are not easily enfolded into the imagined world, and the clarity of the fold line belies Trollope's industrial approach to writing. Although Trollope's autobiography suggests a very objective and dispassionate way of writing that varied little within and between his different novels, this was not, as the composition of *The*

Bertrams attests, always the case. The place of writing, the place of narrative action and the place of reception could fuse to create a troubling challenge to normal practice.

A final reason for Trollope's two falls may lie in the spatial requirements of the story's action. If we turn to the two places where Trollope fell they occur at a similar point within each volume: page 115 in volume one and page 114 in volume two. Given that Trollope wrote 260 words per page he stumbled, in both cases, around 29,500 words, or one-third of the way through each volume. The fact that both these pauses come just after a turning point, or node, within the story suggests that Trollope imagined his narratives spatially; that is, particular things had to happen at particular places within the material text. This suggests that as well as narrative space being an active component of the novel (Crang 2008; Hones 2014), so too is the arrangement of the narrative within the material space of the text. As well as offering another reason for why Trollope enumerated his writing practice, it demonstrates how narrative space was manipulated by material space. All texts, as McGann (1991) observes, have a spatial structure. This can be quite overt, calling attention to itself as in Ezra Pound's richly illustrated Cantos, or it can be more elemental operating at the level of sentence, paragraph or chapter. It can also be instigated by author as opposed to publisher, or by conventions of the time (see Hones 2011, for a discussion of similar themes in relation to Henry James). The spatial arrangement of *The Bertrams* is likely an outcome of all three: Trollope's recognition of audience expectation, his desire to maintain a good relationship with his publisher and his awareness of the continued importance of the triple-decker form. A cursory look at *Doctor Thorne*, written immediately before *The Bertrams*, indicates that Trollope had done something similar in his arrangement of this story. It, like *The Bertrams*, was comprised of 47 chapters, with volume one concluding with Chapter 14 and volume two with Chapter 30. Page 114 equates to page 81 that falls part way through Chapter 6 'Frank Gresham's Early Loves' within the Penguin Classic edition[6]. The reader already knows that Frank most likely loves Mary Thorne, but in this chapter they learn of the scheme to marry Frank off to a rich heiress, thus setting up a doubled plot. Thus, it can be argued that page 114 is what Harriet Hawkins (2015) terms, a place of gathering or Tim Ingold (2011) a knot within a meshwork of interwoven stories; the story arrives here and from hence flows outwards in a new and different way. In *The Bertrams* though, this gathering or knot was at first too tightly woven to permit the story's onward movement. Those whose practice of writing was more psychogenetic might have been able to move on to other moments or parts of the story's composition, but for Trollope whose writing developed logically a conceptual knot required pondering and probing before it was loose enough to proceed.

Notes

1 Most biographers consider that *The Bertrams* was completed on 20 December 1858. This is the last complete week on the work plan. However, close scrutiny reveals that Trollope continued to record his work on this novel through a process of double-entry

recording; sub-dividing weeks on the first page to accommodate the fact that he overran his intended trajectory.

2 See Chapter 3, 'Posting overseas', for greater discussion of Trollope's imagined audience.

3 Although Trollope did write in a relatively linear manner, Mary Hamer (1987) points out that this did not preclude Trollope from returning to his work and re-writing or re-ordering it. In *Barchester Towers*, for instance, Trollope completed volume II on 23 September 1856. The following day he wrote the first four pages of the final volume and then, on 25 September he returned to the ending of volume II to add three further pages before reverting back to pages five and six of the final volume.

4 *The Bertrams* follows the fortunes of George Bertram (the presumed heir to his uncle Bertram's wealth) and, to a lesser extent, his friend Arthur Wilkinson. The pair study together at Oxford and while Arthur goes into the Church, George decides to take a tour of the Holy Lands before deciding whether to follow Arthur into a religious calling. Whilst in the Holy Lands George falls in with his father, Sir Lionel, who he has not seen since he was a child. His father proves completely uninterested in him, desiring his company only for the money he thinks he can sponge from George. On a happier note, George meets, falls in love with and becomes engaged to Caroline Waddington. Caroline refuses to marry George unless he has money, leading to his decision to study for the Bar. Two years go by in which time George's financial prospects do not improve greatly and eventually the couple decide to call off their engagement. Caroline hastily marries the up-and-coming barrister Henry Harcourt, only to discover that he has married her for the inheritance he thinks she will receive. On learning this, Caroline leaves Harcourt only for his success to start crumbling. Crippled by debt he kills himself. Caroline and George are reunited and eventually marry. Meanwhile, Arthur is expected to take over his father's position as vicar of Hurst Staple, but on the understanding that he take a curate's income and gives the remainder of his income to his mother who comes to see the income and the vicarage as her own (his father having died). This situation precludes Arthur from marrying his childhood sweetheart Adela Gauntlet. As George's story unfolds, Arthur battles with his mother to regain his authority and secure the hand of Adela.

5 The triple-decker novel was by this time an established publishing form. See Peter Gaskell (1979) for greater discussion of the publishing norms surrounding the three-decker novel.

6 This calculation accounts for the page breaks in this edition, with each chapter beginning on a new page rather than directly after the end of the previous chapter. It also takes account of the large heading space accorded to each chapter title.

References

Anon. 1859. The Bertrams. *Athenaeum*, (26 March 1859), pp.420.

Anon. 1859b. The Bertrams. *The Saturday Review*, (27 March 1859), pp.368–369.

Anon. 1859c. The Bertrams. *Bentley's Quarterly Review*, (July 1859), pp.456–462.

Anon. 1883. Mr. Trollope as critic. *The Spectator*, (27 October 1883), pp.9–10.

Anon. 1883b. Anthony Trollope. *Macmillan's Magazine*, (November 1883), pp.47–56.

Bushell, S. 2009. *Text as Process: Creative Composition in Wordsworth, Tennyson, and Dickinson*. London: University of Virginia Press.

Couch, D. 2010. *Flirting with Space: Journeys and Creativity*, Farnham: Ashgate.

Crang, M. 2008. Placing stories, performing places: spatiality in Joyce and Austen. *Anglia: Journal of English Philology*, 126 (2), pp.312–329.

Doel, M. 1999. *Poststructuralist Geographies: The Diabolical Art of Spatial Science*. Edinburgh: Edinburgh University Press.

Foucault, M. 1982. The subject and power. In H.L. Dreyfus and P. Rabinow, eds. *Beyond Structuralism and Hermeneutics*. Chicago: University of Chicago Press, pp.208–226.

Gaskell, P. 1979. *A New Introduction to Bibliography*. Oxford: Oxford University Press.

Genette, G. 1997. *Paratexts: Thresholds of Interpretation*. Cambridge: Cambridge University Press.

Gettmann, R. 1960. *A Victorian Publishers: A Study of the Bentley Papers*. Cambridge: Cambridge University Press.

Hamer, M. 1987. *Writing by Numbers: Trollope's Serial Fiction*. Cambridge: Cambridge University Press.

Hawkins, H. 2015. Creative geographic methods: knowing, representing, intervening. On composing place and page. *Cultural Geographies*, 22 (2), pp.247–268.

Hones, S. 2014. *Literary Geographies*. London: Palgrave Macmillan.

Hones, S. 2011. Literary geography: the novel as a spatial event. In S. Daniels, D. DeLyser, J.N. Entrikin and D. Richardson, eds. *Envisioning Landscapes, Making Worlds: Geography and the Humanities*. London: Routledge, pp.247–255.

Hurlebusch, K. 2000. Understanding the author's compositional method: prolegomenon to hermeneutics of genetic writing. *TEXT* 13, pp.55–101.

Ingold, T. 2011. *Being Alive: Essays on Movement, Knowledge and Description*, Abingdon: Routledge.

Levine, C. 2009. Narrative networks: Bleak House and the affordances of form. *Novel: A Forum on Fiction*, 42 (3), pp.517–523.

McGann, J. 1991. *The Textual Condition*. Princeton: Princeton University Press.

Mayhew, R. 2007. Materialist hermeneutics, textuality and the history of geography: print spaces in British geography, c.1500–1900. *Journal of Historical Geography*, 33 (3), pp.466–488.

Moretti, F. 2006. Serious century. In F. Moretti, ed. *The Novel: History, Geography, and Culture, Vol.I*. Oxford: Princeton University Press, pp.364–400.

Russell Perkin, J. 2009. *Theology and the Victorian Novel*. London: McGill-Queen's University Press.

Sadleir, M. 1945. *Trollope: A Commentary*. London: Constable and Company.

Sherman, S. 1996. *Telling Time: Clock, Diaries and English Diurnal Form, 1660–1785*. London: Chicago University Press.

Skilton, D. 1993. Introduction. In A. Trollope *The Bertrams*. London: Folio Society.

Skilton, D. 1972. *Anthony Trollope and his Contemporaries: A Study in the Theory and Conventions of mid-Victorian Fiction*. London: Longman.

Sutherland, J. 1982. Trollope at work on The Way We Live Now. *Nineteenth-Century Fiction*, 37 (3), pp.472–93

Trollope, A. [1883] 1980. *An Autobiography*. M. Sadleir and F. Page, eds. Oxford: Oxford University Press.

Trollope, A. [1875] 1993. *The Way We Live Now*. London: Penguin.

Trollope, A. [1874] 2008. *Lady Anna*. S. Orgel, ed. Oxford: Oxford University Press.

Trollope, A. [1874] 2000. *Phineas Redux*. J.C. Whale, ed. Oxford: Oxford University Press.

Trollope, A. [1859] 1991. *The Bertrams*. G. Harvey, ed. Oxford: Oxford University Press.

Trollope, A. [1858] 1992. *The Three Clerks*. N. John Hall, ed. London: Folio Society.

Trollope, A. 1858. *Work plan for The Bertrams*. MSs. Don c. 9, no.53. Oxford: Bodleian Library

Trollope, A. [1858b] 1991. *Doctor Thorne*. R. Rendell, ed. London: Penguin.

Trollope, A. [1857] 1994. *Barchester Towers*. R. Gilmour, ed. London: Penguin.

6 Writing-through

Making *The Man of Property*

This chapter is concerned with Galsworthy's making of *The Man of Property* (1906): a difficult novel that dealt with themes of marital rape, megalomania and divorce, topics that were still not considered wholly appropriate subject matter for turn of the century audiences. Yet perhaps because of its unconventionality, it was the first of Galsworthy's novels to receive widespread critical attention. It is also the novel that Galsworthy's biographers have seen as his most autobiographical, taking inspiration from his wife Ada's unhappy first marriage and the problems this posed to their own courtship. Although ostensibly fiction, the repeated parallels that Galsworthy's biographers have drawn between the real Ada and the fictional Irene, between Ada's first husband Arthur Galsworthy[1] and Soames Forsyte and the signal happenings within the novel, draw this chapter towards the process of translation from the lived and the visceral to the imagined: how does one spill into the other? This may well ring alarm bells for some; for geographers it could be taken as a return to the work of H.C. Darby (1948) and E.W. Gilbert (1960) and for literary scholars a search for authorial intentions or that elusive thing called inspiration. It is not, however, the purpose or intention to adjudicate on the degree of verisimilitude between one world and another or to search for textual meaning within biography; rather, it is to probe the texture and significance of a writer's lived experience to the development of their creative imagination. Thus, rather than side-stepping the relationship between the real world and the imagined world, this chapter asks how and where are they enfolded within one another?

In so doing, this chapter follows recent work in literary studies and geography that has sought to examine those aspects of composition and creativity that resist and elude rational accounting (Attridge 2004). Broadly, and in line with the overall ethos of this book, this is a concern with process, with the things that happen downstream of the published text, but more precisely it is, as Brace and Johns-Putra (2010) observe, the relationship between these non-representational processes and their representation. Thus, instead of charting instances of the real world within the imagined world, we turn our attention to the lived processes that give birth to these worlds. What this can mean for literary study is that instead of examining the influence of the Potteries on Arnold Bennett's choice of narrative subject and setting for example, we turn our attention to the process (and problematic) of what John Wylie terms, 'writing through' (2010: 49). This is the practice of writing the world in the moment of its experience: it is a simultaneous 'going forth' within, and 'bringing in', of the world. Of course, writing-through may be equally resistant to rational accounting, for 'can there be a writing of sufficient acuity, detail, texture, piety even, so as to perform the work of the "good step", the step that is adequate to the ground it clears? (Wylie 2010: 60). This implies that any writing will always fail to capture and express experience, but more than this, it suggests that any practice of writing is always a negotiation. Hence, within any attempt to write-through the world there is always an inescapable absence, a gap or breach between the world outside and the world inside, between what is near and what is far, with the result that writing-*through* often becomes writing *of*. Where Wylie turns to the philosophical writings of Jean-Luc Nancy in order to disturb these binaries, they can be agitated, or at least complicated, if we alter the focus of writing-through.

Wylie's interest in writing-through centres on landscape writing as a particular, albeit largely overlooked, literary genre[2]. It is a genre that draws from regional writing and travel writing, where sense of place and geographical perspective are critical to both the process and product of writing. What happens, however, if process and product are decoupled, the geographical inflection of genre – landscape, region, travel – is left aside and process comes to the fore? This decoupling turns our attention from the accuracy of the 'good step' to its performance: to what happens as the heel hits the ground, the body rocks over and the weight of the self rolls from heel to toe. Although analogy can only take us so far, it points to how writing-through can enable us to attend to what goes on along the journey from world to word: the process of experience, translation and expression. It encourages us to ask, following Brace and Johns-Putra (2010), what are writers moved to represent, but also, how do they choose to represent that which has moved them? Elsewhere we might term this the process of storytelling or narrativisation which, according to Hayden White, is the process 'of fashioning human experience into a form assimilable to structures of meaning that are generally human rather than culture-specific' (1980: 5). It is, therefore, the process of translating self-experience into social experience or, put another way, that which is unique into that which is shared. Most frequently, literary geographers examine this at a formal level, utilising close-reading to explore how narrative strategies, such as point-of-view, time, distance, and space generate affects within and beyond the novel (Crang 2008; Hones 2014). What, though, of the making and development of these strategies within the compositional process? Where Hones (2014) pushes geographers to read literature in more spatially nuanced ways, this chapter applies this thinking to the compositional process. It brings together ideas of writing-through with the close-reading strategies of literary geography to suggest that both can illuminate the things that happen before the text.

Writing as therapy: *The Man of Property*

Today creative writing is frequently used as a therapeutic tool, a way to aid or further physical, mental or emotional healing and promote wellbeing. According to James Pennebaker (1989) creative writing serves two therapeutic functions. On the one hand, it helps us articulate thoughts and feelings that would otherwise remain repressed, thereby relieving stresses and pressures that may incapacitate us. Writing is a slower process than speech and therefore it gives us time and space to tease out the complexities of our subjective experience. On the other hand, creative writing enables us to bring order, sense and coherence to traumatic experiences; the process of narration, both in terms of the time it takes and the demand it places on the individual to externalise their experience and emotions, reduces the immensity of experience making it more comprehensible. In this way, creative writing becomes a means, following Sigmund Freud (1914), of working through: of going over events and memories multiple times so that they cease to exert so strong an emotional hold upon us, so that we can see them from other perspectives and so that we overcome our resistance to fully acknowledging that they happened. Creative writing also, according to Michael White and David Epston (1990), serves as a way of challenging culturally sanctioned stories or dominant ways of knowing. It allows people to tell counter-stories and reframe accepted ways of knowing the world. It is therefore, a practice that is potentially disruptive and unsettling, not necessarily for the self, but in its social effects and affects.

Obviously, such thinking is of recent origin, but the relationship between creative writing, self-experience and therapy does reach back much further, albeit without the therapeutic label. Sigmund Freud was a great admirer of John Galsworthy, referencing, in both *The Psychopathology of Everyday Life* (1904) and *Civilisation and its Discontents* (1930), his psychological sensitivity and capacity to naturalise, articulate and rationalise the unconscious (Langbauer 1999). Freud was evidently praising Galsworthy's represented world but this, as Galsworthy observed, was not a world separate from the lived world:

> unless you are an absolute genius, you must look near to home, for the material you work with . . . It is a sort of cant to believe that the artist can reproduce anything; this with the exception of a few great geniuses is not so; the artist can as a rule only produce that, or things similar to that, which he [sic] has felt. To go deep you must feel deeply, on your own lines whatever they may be. I am myself no believer in barring out the personal or what may be construed into the personal theme.
>
> (11 September 1905, Marrot 1936: 182)

Galsworthy made this observation in a letter to his sister, Lilian, prior to the publication of his novel *The Man of Property*. Galsworthy had sent Lilian a copy of the novel's final draft and while Lilian's letter no longer survives we know from the rationalising tone of Galsworthy's reply that she was troubled by the similarities between the worlds of the Forsytes and the Galsworthys, so much so that she asked Galsworthy not to publish it.

Set in 1886, although written in the early years of the twentieth century and published in 1906, *The Man of Property* was a critique of Victorian materialist values; it was a novel of possession that examined the notion of ownership at the heart of Victorian society. It followed the sprawling, upper-class Forsyte family: a self-made, professional family for whom property equated with success and prestige could be bought. At the heart of the novel was the troubled marriage of Soames and Irene Forsyte. As the man of property, Soames' life was indexed in terms of his possessions and their worth and of course, his most prized possession and acquisition of greatest value was his wife, Irene. Galsworthy used this marriage to explore the meaning of property and possession within the domestic realm and the bounds of what was permissible in the exercising of this possession within both home and marriage. Lilian's concerns were focused on the apparent similarities between Galsworthy's soon-to-be wife Ada and the fictional Irene – particularly given Ada's unhappy first marriage –, on the familial resemblances between the Forsyte family and the Galsworthy family and on the rape scene that lay at the heart of the novel[3].

The autobiographical strain of this novel has been iterated many times (Edwards 2011; Gindin 1987; Pallette 1959), yet what has received less attention is how this novel emerges from, rather than expresses, biographical experience. Thus, in an echo of Noel Castree and Thomas Macmillan (2004) and Brace and Johns-Putra (2010), there is a need to study the practice of representation, which is, in this instance, 'how' Galsworthy turned the lived world into the imagined world and the translations, distortions and absences this entailed. Felt or lived experience was a necessary precursor to Galsworthy's creativity. Several years before he began writing he disclosed the wish that he 'had the gift of writing . . . only it isn't really the writing so much as the thoughts that one wants; and, when you feel like a very shallow pond, with no nice cool deep pools with queer and pleasant things at the bottom, what's the good?' (8 September 1894, Marrot 1936: 97). Galsworthy was 27 and there had been little in his life up until this point that had aroused any deep or passionate feelings. His life was one of comfort and security, and this, as he suggested to the letter's recipient, Monica Sanderson, was not conducive to creativity[4]. The relationship between felt experience and writing was one Galsworthy elaborated upon in a lecture on character. In this, he stressed how 'a real incident, or person, impinges sharply on a receptive mood of a novelist's particular nature or temperament'. There gathers around this similar 'impacts or impressions that have been stored in the sub-conscious mind' until these swell 'to proportions which demand the relief of expression, and in written words the novelist proceeds to free himself [sic] (Galsworthy 1933: 272). Galsworthy's choice of language in this closing line is interesting, for in suggesting that the self is imprisoned by experience it anticipates many of the debates over creative writing's therapeutic potential. The intimation is that creative writing, flowing as it does from felt experience, performs an emancipatory or cathartic function: it has the power to tease out and give voice to the tangle of pent up experiences and emotions.

As several of Galsworthy's biographers have remarked, Ada was Galsworthy's defining experience and it was she, and her cruel treatment by society, that made him feel (Barker 1963; Gindin 1987). As James Gindin (1987) observes, Ada became Galsworthy's locus of self-definition; it was her lived experiences that were to free him from the strictures of his comfortable social world, and which were to inspire his creative writing from the turn of the twentieth century through to the outbreak of the First World War. *The Man of Property* though, was not, as Galsworthy observed in his reply to Lilian, a direct copy of reality; something happened *en route* from world to word. As he explained, his artistic temperament was like that of a 'chemist, more cold, more dissective, always riding a philosophical idea, and perverting, if you like, my values to fit' (11 September 1905, Marrot 1936: 181). This has been taken as an interest in the machination of things rather than with the thing itself; for instance, rather than seeking to defeat the property instinct within *The Man of Property*, Galsworthy preferred to explore the workings of property (Gindin 1987). This analogy though, also returns us to the idea of writing-through, or working-through if we return to the language of therapeutic writing, for it suggests that a process of distillation lay at the heart of Galsworthy's writing practice. Thus, although Galsworthy folded his life into the creative process it was a folding that was more akin to origami than enveloping; there were traces of lived experience but they had been folded in ways complex and distorting. Thus, exploring *The Man of Property* in relation to writing-through is not about linking book and biography, but rather, about exploring the decisions, possibilities, agencies and histories that converged in its making.

A modernist empty space: the making of Irene

At the centre of Joseph Conrad's novel, *The Heart of Darkness* (1899), there is an absence. The heart is never reached, Kurtz's story, what actually happened to him, is never told. As Daniel Just observes, following Northrop Frye, it is a dislocated story, one that 'oscillates around a central situation that never becomes a direct object of the story' (2008: 276). We know of Kurtz, but we never know him; rather, he and his story haunt the narrative as fragments gather and then disperse never coalescing into anything tangible. Haunt is perhaps the wrong word, for it implies that Kurtz is a spectral presence, a person or thing that to follow Jacques Derrida (1994) is revenant – it has returned. Kurtz however, never went away but nor is he present; instead, he is an absent presence, we may feel him, touch him at times, even hear him and see him, but we never get close enough or intimate enough to understand him. *The Man of Property*, it can be suggested, plays with similar ideas of absence, albeit not with the same degree of poise.

The absence at the centre of *The Man of Property* is Irene Forsyte, Soames' wife, and the character who was most troubling to Galsworthy's sister, Lilian. Lilian's principle anxiety regarding *The Man of Property* was what Galsworthy termed the 'personal question'. Namely the similarities she felt existed between the Galsworthys and the Forsytes and, most particularly between Ada Galsworthy and Irene Forsyte. In an attempt to dispel her concerns, Galsworthy observed, 'who really knows enough or takes enough interest in us to make it more than a two days wonder that I should choose such a subject' and who 'knows enough even to connect A with I, especially as I have changed her hair to gold' (11 September 1905, in Marrot 1936: 182). The last line may seem a little flippant but, as Galsworthy wrote, his friends, the Garnetts, the Conrads and the Hueffers did not care and as for family, given recent events they were unlikely to read the novel. Despite the similarities between the real Ada and the imagined Irene, what is most interesting about the 'figure of Irene' is that, as Galsworthy observed in the 1922 preface to the first volume of the Forsyte saga, she is '. . . never, as the reader may possibly have observed, present, except through the senses of other characters' (1922: x). In consequence, and perhaps taking inspiration from Joseph Conrad, Lynne Hapgood (2000b) has described Irene as a modernist empty space at the centre of the narrative. Like Kurtz in *Heart of Darkness*, she is, in many ways, the lynchpin of the narrative, but yet we never know her, she is an absence, but one that presses in upon the narrative.

The reader first encounters Irene in the opening chapter where she is described as 'A tall woman, with a beautiful figure, which some member of the family had once compared to a heathen goddess . . . Her hands, gloved in French grey, were crossed one over the other, her grave, charming face held to one side, and the eyes of all men near were fastened on it' (2001: 17). As the chapter closes we overhear Irene talked about as 'a good-lookin' woman', 'a pretty woman' and 'distinguished-looking' (2001: 27–28). Soames describes her as 'ever silent, passive, gracefully averse' (71), while to Soames' cousin Jolyon, Irene has a face of 'soft, persuasive passivity' (253). Through each of these perspectives the reader catches a glimpse of Irene, but the narrative never stays with her and while we may occasionally hear her, we never see the world from her perspective. Instead, we get a cubist-esque figure: a fractured, partial and elusive Irene. Thus, it is not that Irene is absent *per se*, she is a visible figure and she figures within the unfolding narratives, she has bodily presence; rather, she is more an absent presence, for she does not inhabit the space or role we expect of her.

Where Hapgood focuses on Irene as a measure of the modernist sensibilities of the text, Rudolf Sauter has suggested that Irene's vagueness was not aesthetically motivated as such; rather, one rarely sees her 'in the round' because Galsworthy was 'at pains to say: "she shall *not* be a picture of Ada as a young woman"' (1967: 79). Irene's emptiness then, could be construed as a way of distancing the real from the imagined. As writing strategies go it was not that successful, for along with Lilian Sauter, the Garnetts linked Irene to Ada, but the intriguing point comes a little earlier in Rudolf Sauter's recollection, when he acknowledges that Ada was undoubtedly Galsworthy's inspiration for Irene, but that in the 'working out' (1967: 78) of this original idea it got deformed and reformed many times.

To understand the nature of this 'working out' it is useful to turn once more to John Wylie's reflections on the relationship between landscape and writing, where sometimes 'Proximity induces not familiarity and recognition (let alone fusion); instead it gives life to a distance and an estrangement. The closer you get, the more an unbridgeable gap seems to impose itself' (2010: 57). In journeying towards something, some place or someone it, or they, can, instead of becoming clearer, recede from view. It becomes strange or to echo Wylie, uncanny as previously unseen, unheard or unappreciated parts come to the fore. Thus, what one thought one knew, turns into something unsettling and unknown. To fold this back into the creative process, Charlotte Doyle (1998) has argued that after the first flash of inspiration, or seed moment, writers often enter what she terms, the 'fictionrealm'. This is the world of the narrative and it is here where characters begin to take on a life of their own, transforming from an idea into a more complex, living entity with a motive force of their own. In his response to Lilian, Galsworthy made a very similar observation:

> As to character there is always this to be remembered. You start with a suggestion, you go on working from a figure (a living figure) for perhaps two or three chapters, then suddenly you work no longer from that figure, but from what you have said about him – from your own creation in fact, which at every sentence diverges more and more from the original.
>
> (11 September 1905, in Marrot 1936: 184)

Galsworthy was making direct reference to his and Lilian's father here, but it is a sentiment that seems to apply as much to the link between Ada and Irene. The seed of Irene may well have begun in Ada's social defiance, as 'a women rebelling against the fetters of chattelhood' (Sauter 1967: 78), but as this seed germinated and grew its resemblance to Ada became distorted, with Irene becoming, in Sauter's words, 'a symbol, an evocation of beauty, impinging upon a materialistic world' (1967: 78–9).

Galsworthy's intimation is that this process was quite spontaneous, but one function of the author, as Viktor Shklovskij ([1917] 1986) observes, is to make us look anew at the world and one way of doing this is through a strategy of defamiliarisation. This is a way of writing the world so that we overcome our automatic and habitual ways of seeing and knowing it. By making the familiar strange – by presenting objects in unusual settings, describing them as if seen for the very first time, or offering only partial glimpses – Shklovskij believed that we could develop a fuller sense of perception and a deeper, more critical awareness of the world. Although neither common nor familiar, Ada's situation was, in a sense, unwriteable. It is worth noting that many of Galsworthy's contemporaries struggled in their representation of women. H.G. Wells, Arnold Bennett and Ford Madox Ford all propounded progressive attitudes towards women and marriage in their work but often concluded with something more regressive that reaffirmed contemporary and accepted visions of women as wives and mothers (Flanagan 2000/1). Undoubtedly aware of the horizon of expectation (Gadamer 1975), Galsworthy's working-out of Irene as a character that is not 'vitalized as the Forsytes are vitalized' (11 September 1905, in Marrot 1936: 183), and so hovers on the margins of the novel, may well have been both pragmatic and progressive. A powerful female character who narrated the living through of chattelhood was unlikely to win sympathy with the Edwardian reading public. A more ambiguous and fleeting character, one whose story we are not certain of, does not determine a response in quite the same way; instead, Irene suggests that there is another narrative of womanhood that awaits writing. As Hapgood (2000b) suggests, Irene symbolises the future; she is a new kind of woman, but one it is impossible to realise in the world she inhabits or, perhaps, to articulate in the world Galsworthy inhabits. Thus, absence works to distance 'A from I', but it also works to distance Irene from the vicissitudes of real life and thereby makes it more possible to generate sympathy for Irene's situation and to subtly disrupt, or defamiliarise, Edwardian notions of femininity.

Writing uncertainty: distancing the real and the imagined

Absence is a distancing strategy, a way of differentiating and separating the real from the imagined. At the same time though, it is a method or technique that points to Galsworthy's immersion within a particular literary moment. Around the time Galsworthy was crafting *The Man of Property,* Joseph Conrad and Ford Madox Ford were experimenting with new ways of writing the world and they, along with Henry James, came to be associated with literary impressionism and its quest to render the immediacy of the moment. There is much disagreement over what literary impressionism is and what or who falls under its mantle (Peters 2001; Matz 2001). Ford Madox Ford's *On Impressionism* was the first real articulation of it as an aesthetic style and, as Max Saunders observes, it was temporally and spatially broad, tracing a 'line of technical self-consciousness . . . from Flaubert through Henry James and into the formal experiments practised by Conrad, Ford and Pound' (2004: 23). Thus, it spilled into realism at one end and modernism at the other, and incorporated writers like Guy de Maupassant, Gustav Flaubert and Ivan Turgenev[5]. Galsworthy was a long admirer of these writers and wrote on various occasions of his attempt to emulate their style (1919; 1928), yet he has rarely, if ever, been considered an impressionist writer. In light of Galsworthy's entire oeuvre this is perhaps a valid evaluation, but in relation to *The Man of Property* which, as the *Times Literary Supplement* (TLS) noted, was strongly influenced by the work of these masters, it seems something of an oversight.

According to the TLS the influence of these masters was most notable in the 'constant succession of details' and the 'progression' of points of view, which made 'the progress of the story . . . the progress of your comprehension of those . . . points of view' (Anon. 1906). It observed for instance, that the reader did not move with, or gain insight into the point of view of Bosinney or Irene; rather, it was with the views of the Forsyte family that one progressed. It is interesting that the TLS focused on Galsworthy's method more than it did the story, for in Ford's view, impressionism was more a technique than an aesthetic movement. There were significant and experimental practitioners but there were also those who engaged with the technique in more modest, pragmatic and discontinuous ways. Galsworthy, it can be argued, was one of these, reflecting that a writer 'is by no means tied' to any one technique. A realist, for instance, 'may be poetic, idealistic, fantastic, impressionistic' for the word 'characterises the artist whose temperamental preoccupation is with revelation of the actual inter-relating spirit of life, character and thought' (1913: 272–3). Moreover though, an overriding feature of literary impressionism, argues Paul Armstrong (2015), is the manner in which it calls attention to its method of construction. In seeking to render the 'odd vibration that scenes in real life really have' (Ford 1914: 175), impressionist writers play with the conventions of realism in order to resist the building of pattern or story. Thus, what is made present draws our attention to those things that are absent and which we expected to find within the story and this, as Armstrong observes, is the paradox of impressionism for the way absence and presence intertwine within fiction draws the reader's attention to its process of representation: to the formal properties of the work. The question that suggests itself then, is did Galsworthy draw upon strategies of absence to emphasise the fictiveness of *The Man of Property* and distance it from the real?

To answer this, let us return to Galsworthy's concern with multiple perspectives, which as we have seen in relation to Irene, present the reader with a fractured and partial sense of her. If we examine these perceptions of Irene what becomes apparent is that they alter over time. In the opening few chapters we know, through several Forsytes, of Irene's beauty, but as the novel develops past and present mingle offering a different vision of Irene. In 'Progress of the house', the first chapter of part two and one that seemingly has a forward propulsion as Soames considers the development of his suburban retreat, Robin Hill, the reader is also made privy to the memory of his courtship of Irene; to his 'love at first sight' (2001: 112) to her resistance and adversity, and to his promise that if their marriage should be a failure he would set her free. Irene as passive object is here recalled as an agent with a mind of her own; self-assured, assertive and pragmatic. As part two draws to a close yet another Irene emerges, as Irene the desirable object gives way to Irene the unruly wife in need, Soames believes, of a 'good beating' (2001: 228). Soames' oscillating perceptions of Irene disrupt what Saunders (2004) terms, the chronological straightforwardness associated with realism. More than this though, and following Saunders further, the oscillation of perspectives tells us something about the nature of perceiving and knowing.

Realism might give us illusion and then disillusion as the image is gradually shattered by the reality and one vision is replaced by another, yet how we see and how we know the world rarely works like that. It is more likely that we hold competing visions in tension with one another. Thus, as knowledge of another grows, new information does not delete old; instead, the original image becomes richer, more complex and more confusing as discontinuous ideas co-exist. At the close of *The Man of Property*, when, after Bosinney's death, Irene returns to Soames, we see this uneasy coexistence played out:

> he longed to cry: 'Take your hated body, that I love, out of my house! Take away that pitiful white face, so cruel and soft – before I crush it. Get out of my sight never let me see you again!'
>
> And, at those unspoken words, he seemed to see her rise and move away . . . without a thought of him, without so much as the knowledge of his presence.
>
> Then he cried, contradicting what he had not yet spoken, 'No; stay there!'
>
> (2001: 312)

Soames knew by this time that Irene had most certainly been Bosinney's lover, but this is knowledge he has discerned and not been told; he is certain and yet uncertain, wavering between love and hate, cruelty and compassion, anger and composure. This is not a chapter of closure but of bewilderment, disorder and disorientation. These were all strategies common to impressionist writers and were used to disrupt the smooth unfolding of the narrative, to emphasise the provisionality of knowing and to upturn any sense of a stable totality (Armstrong 1983). Perhaps though, Galsworthy closed with bewilderment not only to convey epistemological uncertainty but, as Saunders observes in relation to the work of Ford Madox Ford, to say something about the feelings that pulsate within the epistemological process. Impressionism tends to focus on perception, on the seen world and how this effects our knowing, rather than on the felt world, yet as Saunders writes, passion, 'as both love and suffering, is the hallmark of the Fordian impression. For it is passion which makes the deepest impression, or allows other things to make their impression under its sign. After all, disillusion is only painful when we care about the illusions' (2004: 46). Saunders is referring to Ford's 1915 novel *The Good Soldier* in which Ford renders the strange mixture of contrary feelings we always hold towards others; love lingers even when confronted with infidelity. We know from Galsworthy's letter to his sister Lilian that he did not like 'a contrast so given as to enlist the reader on one side, and against the other' (11 September 1905, in Marrot 1936: 181) for character was not reducible to good or bad, hero or villain. Trying to express this lived reality on the page was to recognise the limits of realism and to appreciate that putting distance between the real and the imagined did not necessarily mean bending the former out of all recognition. Instead, distance could be created by being truer to the complexities of the real than realism permitted, or, put another way, of showing our human struggles not solutions. In this context, writing-through is about capturing, rather than neutering, the emotional unsettledness of lived experience.

Playing the long game: defeating the Forsytes

Sarah Edwards (2011) has argued for *The Man of Property*, and the Forsyte saga more broadly, as a neo-Victorian novel. By this she means that it tracks, often quite self-consciously, between identification with and rejection of Victorian values. Most notably, it takes marriage, so frequently the narrative closure for many Victorian novels, and begins to dismantle it, exposing the patriarchal power and possessive instincts that sustain it. It is, Edwards notes, a divorce novel; it was written at a time when reform of Britain's Victorian divorce laws was being debated and when Galsworthy was living through the inequalities of these laws in his own relationship with Ada. Galsworthy and Ada met in 1895 but did not marry until 1905. The reason for their ten year courtship was that Ada was already married to Galsworthy's cousin, Arthur. Little, barring dates, is known about Ada's first marriage. We know she married Arthur in 1891, ceased to live with him in 1902 and initiated divorce proceedings in December 1904. The reasons for their marital disharmony are uncertain although, as Gindin observes, all sorts of things have been imputed, not least the violence of the fictional marriage of Soames and Irene, which has been folded back on to that of Ada and Arthur. That Ada's first marriage in some way influenced Galsworthy's conception of *The Man of Property* seems likely, but in transposing lived experience to the page Galsworthy battled with how to generate sympathy for the 'illicit love' (1 June 1905, in Marrot 1936: 169) that lay at the heart of the novel.

The illicit love concerned Irene Forsyte and Philip Bosinney, an architect and, when the novel opens, the young man engaged to Soames' niece, June Forsyte. To embark on a love affair may seem at odds with what we know of Irene, but in keeping with the spirit of her character it is a quiet affair glimpsed by others but never fully understood. One such observer was Swithin Forsyte, Soames' uncle, who saw Irene's 'soft, warm body clutched, her head bent back from his lips; his kiss; her recoil; his cry: 'You must know I love you" '. Soon after though, 'Swithin awoke' (2001: 129–30), questioning the veracity of what he saw or dreamed. In an early draft, sent to Edward Garnett, Galsworthy's way of dealing with this affair, was for Bosinney to commit suicide after learning that Soames had raped Irene, and for Irene, in the last chapter, to return to Soames crushed and defeated. Neither of these, in Garnett's view, was plausible. Bosinney's suicide, in particular, was '*an artistic blot* of a very grave order . . . When I read it, I said, "incredible", "not in keeping with what we have learnt of B" '. Moreover, 'I think the last chapter is better away. I don't think it strengthens but weakens the final effects. It *talks on*-after the curtain has dropped' (27 May 1905, in Marrot 1936: 165–166, emphasis in original). A far more plausible ending, Garnett argued, would be for Bosinney 'with £20, borrowed or begged' to 'have started with Irene and her jewels for Paris . . . The young fresh forces of love and life are on *their* side: to make them throw up the sponge suddenly is an *artistic débâcle*' (27 May 1905, in Marrot 1936: 165).

Galsworthy came round to Garnett's way of thinking with regard to Bosinnney, agreeing that:

> the suicide would most probably have come about under the wheels of a 'bus in the fog, instead of the following afternoon. I can alter this, with but little change; and if you like I can leave it in the minds of the reader as it would be in the minds of the Jury, doubtful whether it was suicide or no.
>
> (2 June 1905, in Marrot 1936: 170)

Over the ending though, Galsworthy was far more intransigent believing the final chapter, where Irene returns to Soames, integral to the artistic conception of the work:

> You, and I think your wife, want me to end the book with a palpable and obvious defeat of Forsyteism by making the lovers run away happily.
>
> To my mind (and I desire to defeat Forsyteism) the only way to do so is to leave the Forsytes masters of the field. The only way to enlist the sympathies of readers on the other side, the only way to cap the purpose of the book, which was to leave property as *an empty shell* – is to leave the victory to Soames.
>
> (1 June 1905, in Marrot 1936: 169, emphasis in original)

Such a method of composition was, as Galsworthy observed to his sister, Lilian, a negative one: 'I deal always negatively, destructively, I can't bear the idea of a beautiful character, it seems to me so obvious, commonplace and disgusting, above all eminently unphilosophical' (11 September 1905, in Marrot 1936: 181). We might, after Ford Madox Ford, term this a critical attitude: a desire to unsettle accepted ways of thinking, knowing and doing. Critical writing, wrote Ford, was intended to affect, challenge and disrupt:

> The fact is that new ideas almost invariably affect our sense of values – our sense of the whole values of everything. If you tamper with the ideal of the marriage laws you will interfere not only with the sanctity of the home, but also with the sense of property, with values in literature, and with all ideas of human relationships. To interfere with these ideas is very seriously to diminish the motive force that impels men to any protracted and consistent course of action.
>
> (Ford 1911: 13)

Ford could well have had *The Man of Property* in mind when he wrote this for its negative method was designed to surprise; to offer a version of life that was not expected and which, at a narrative level, left life messy and unfinished rather than neatly folded and tidied away. It also surprised by replacing a linear form – the triumph of love that Garnett petitioned for – with a cyclical one (Miller 1994). By insisting on the necessity of the last chapter wherein Irene returns to Soames, we see not only the defeat of love and youth, but also a symbolic circling as the narrative returns to the place where it began: at home. Yet the homes that bookend the novel could not be more different. The novel opens with an 'at home' to celebrate the engagement of June Forsyte to Philip Bosinney, at which all 'the family were present' (2001: 11). It closes with Soames slamming his front door on this unit. The Forsytes, who appeared at the start in confident 'full plumage' (2001: 11), their home the centre of a well-ordered, finely-tuned society built upon property and ownership, have been left somewhat drift, for the home and all it stands for has been disturbed from within.

According to Jane Miller (1994) the ending of *The Man of Property* subverts formal closure through its thematic openness. The reader is left in a state of 'unknowing': what is going on behind the closed door, will Irene remain with Soames, will Soames undergo an ideological epiphany, and will his belief in property remain sound? This inconclusiveness was purposeful, Miller (1994) argues, for it was the only way Galsworthy could maintain the critical voice of his text. To have offered narrative resolution in the form of a reuniting of Irene and Soames or the running away of Irene and Bosinney, would have been to reaffirm the primacy of marriage and love to both the social order and the fictive form, whereas the novel's thematic openness encourages a sense of discomfort, uncertainty and frustration on the reader's part. This lack of narrative closure or finality was another characteristic of impressionism (Armstrong 1983) and was intended to pique the reader's curiosity, heightening their alertness to the social world and keeping the themes alive beyond the space of the novel. What it also suggests however, was Galsworthy's own uncertainty over how to end *The Man of Property*. It was a bold book for its time, particularly in its desire to draw attention to the damaging consequences of marriage for women. In the main, reviews of the novel were favourable (see Marrot 1936), but there were some, such as that in *The Spectator*, which were uncomfortable with the world Galsworthy had created:

> . . . [It is] so able that it cannot be overlooked, and so ugly in places that it cannot be recommended without serious caution. . . . Mr Galsworthy does not linger over the repellent details which he has thought it necessary to introduce, but though they are handled in a coldly dispassionate manner, their presence lends the book a quality which renders it unacceptable for general reading.
>
> (Marrot 1936:185).

Published in 1906, around the time that society was beginning to grapple with the divisiveness of the Suffragette movement, *The Man of Property* had to chart a tense social politics. To write too bold an ending to *The Man of Property*, to allow illicit love to prevail or to have Irene victorious in independence was too reductive and simplistic. It would leave Irene scorned, it would make obsolete the novel's critical force and, if there was any trace of Galsworthy and Ada's experience in that of Irene and Bosinney, it would underplay the emotional anguish that had characterised their ten-year love affair. The form of the novel, in a way, could not accommodate the complexities of this relationship or its duration. Moreover, a positive ending would write the world in a conventional and expected manner, for as Garnett noted, Bosinney and Irene's elopement was '. . . what I thought was bound to come' (27 May 1905, in Marrot 1936: 166). A happy ending, albeit through marriage, was the standard structure of a realist narrative. It closed discussion and would leave 'the Forsytes victorious with *the reader (general reader)'*. An unexpected and unhappy ending made illicit love more acceptable; it 'leaves the lovers victorious even with them [the reading public], for it is only the beaten who are victorious' (1 June 1905, in Marrot 1936: 169, emphasis in original).

Writing-through

In his meditation on writing-through, Wylie, taking inspiration from the philosophy of Jean Luc Nancy, argues that if landscape is uncanny, both simultaneously strange and familiar, then writing about this landscape is to inhabit a between space that is both inside and outside, near and far, home and away. This means that the 'good step', with which this chapter began, is always an impossibility, for the world we express on the page is never quite the world of our lived experience: things happen, intentionally and unintentionally, between place and page. Although fiction rarely seeks to capture lived experience in the same way as landscape writing, as Galsworthy acknowledges of *The Man of Property*, this was a novel that clearly originated from personal experience, and, it was in this space between place and page that Galsworthy's real and imagined worlds mingled, merged and morphed. As such, this between space, or, to put it another way, the place of making, appears very much as a third space: an 'open-ended set of defining moments' (Soja 1996: 260). Third space is a place of possibility, a place where things, ideas and people co-mingle in the becoming and articulation of something new. Third spaces, then, have radical qualities, often being places where marginal voices are heard, differences enunciated and new ways of being forged.

We see this in the experimental style of *The Man of Property*. In crafting it, Galsworthy began with the real, but as this 'real' filled out and took shape on the page it twisted, turned and transformed. Strategies of effacement and defamiliarisation, impressionistic conventions and elliptical structures became Galsworthy's mechanisms for representing, or more aptly, *re*-presenting the real in a manner that blunted the edges of its nonconformity while still sustaining its progressiveness. The result was something familiar yet strange, known yet unknown, lived and yet not lived. It is impossible to know whether Galsworthy found writing-through his and Ada's experience therapeutic, but its affects were undoubtedly transformational. This was a novel that in both its structure and its narrative strategies presented the world in a new way, a way that was neither realist nor modernist, which differed from the social novels and New Women writing of the 1890s and which spoke in a voice pregnant with radical possibility. As Ford Madox Ford observed, in 'the novel, as in the play, Mr Galsworthy pursues the quality of surprise with a dogged, an unblinking persistency. You never catch him nodding; he never nods' (1911: 88); his worked carved its own groove, but it is a groove that has not, perhaps, been fully appreciated.

Notes

1 Arthur was John Galsworthy's cousin.
2 It can be argued that this is altering given the growing popularity of landscape writers like Robert MacFarlane, Roger Deakin and Heather Macdonald.
3 The plot of *The Man of Property* follows Soames Forsyte as he attempts to own and possess his wife Irene. He builds a house for her, Robin Hill, out in the suburbs in an attempt to control her circle of friends. Irene resists Soames, falling in love with Phillip Bosinney, the architect hired to design Robin Hill (and June Forsyte's fiancée, the daughter of Soames' cousin Jolyon). Distraught to learn that Soames has raped Irene, in an attempt to assert his right of property, Bosinney, wandering in a London fog, is run over and killed, leaving Irene with no option but to return to Soames.
4 Monica Sanderson was the sister of Ted Sanderson, one of Galsworthy's early school friends and later travelling companion on his journeys to Australia and South Africa.
5 As Saunders observes, critics of Ford's *On Impressionism* were uneasy with his criteria for what constituted an impressionist writer. Why, for instance, was Maupassant an Impressionist and not a Realist or a Naturalist? Moreover, why was Ezra Pound an Impressionist and not a Modernist? Ford's broad approach was criticised for thus grouping together writers who were so very different.

References

Anon. 1906. The Man of Property. *Times Literary Supplement*, (30 March 1906), p.116

Armstrong, P. 2015. What is it like to be conscious? Impressionism and the problem of qualia. In G. Castle, ed. *A History of the Modernist Novel*. Cambridge: Cambridge University Press, pp.66–85.

Armstrong, P. 1983. The hermeneutics of literary impressionism: interpretation and reality in James, Conrad and Ford. *The Centennial Review*, 27 (4), pp.244–269.
Attridge, D. 2004. *The singularity of literature.* London: Routledge.
Barker, D. 1963. *John Galsworthy: The Man of Principle*. London: Heinemann.
Brace and Johns-Putra, A. 2010. Recovering inspiration in the spaces of creative writing. *Transactions of the Institute of British Geographers*, 35 (3), pp.399–415.
Castree, N. and Macmillan, T. 2004. Old news: representation and academic novelty. *Environment and Planning A*, 36 (3), pp.469–480.
Crang, M. 2008. Placing stories, performing places: spatiality in Joyce and Austen. *Anglia: Journal of English Philology*, 126 (2), pp.312–329.
Darby, H.C. 1948. The regional geography of Thomas Hardy's Wessex. *Geographical Review,* 38 (3), pp.426–443.
Derrida, J. 1994. Spectres of Marx. *New Left Review*, 1 (205), pp.31–58.
Doyle, C. 1998. The writer tells: the creative process in the writing of literary fiction. *Creativity Research Journal*, 11 (1), pp.29–37.
Edwards, S. 2011. The rise and fall of the Forsytes: from neo-Victorian to neo-Edwardian marriage. In M.-L. Kohlke and C. Gutleben, eds. *Neo-Victorian Families: Gender, Sexual and Cultural Politics.* Amsterdam: Rodopi B.V., pp.197–220.
Flanagan, A.-M. 2000/01. Ford's women: between fact and fiction. *Journal of Modern Literature*, 24 (2), pp.235–249.
Ford, M.F. 1914. On Impressionism. In, H. Munro, ed. *Poetry and Drama, Vol.II*. London: The Poetry Bookshop.
Ford, M.F. 1911. *The Critical Attitude*. London: Duckworth.
Freud, S. [1930] 2002. *Civilisation and its Discontents*. London: Penguin.
Freud, S. 1914. Remembering, repeating and working-through. *Standard Edition*, 12, pp.145–156.
Freud, S. [1904] 1914. *The Psychopathology of Everyday Life*.London: Fisher Unwin. Trans. A.A. Brill.
Gadamer, H.-G. 1975. *Truth and Method.* London: Continuum Books.
Galsworthy, J. 1933. *Candelabra: Selected Essays and Addresses*. London: William Heinemann.
Galsworthy, J. 1928. *Castles in Spain*. London: William Heinemann.
Galsworthy, J. 1922. *The Man of Property*. London: William Heinemann.
Galsworthy, J. 1919. *Another Sheaf.* London: William Heinemann.
Galsworthy, J. 1913. *The Inn of Tranquillity*. New York: Charles Scribner.
Galsworthy, J. [1906] 2001. *The Man of Property*. London: Penguin.
Gilbert, E.W. 1960. The idea of the region. *Geography*, 45 (3), pp.157–175.
Gindin, J. 1987. *John Galsworthy's Life and Art: An Alien's Fortress*. Ann Arbor: The University of Michigan Press.
Hapgood, L. 2000. The unwritten suburbs: defining spaces in John Galsworthy's The Man of Property. In L. Hapgood and N. Paxton, eds. *Outside Modernism: In Pursuit of the English Novel, 1900–1930.* Basingstoke: Macmillan, pp.162–79.
Hones, S. 2014. *Literary Geographies*. London: Palgrave Macmillan.
Just, D. 2008. Between narrative paradigms: Joseph Conrad and the shift from realism to modernism from a genre perspective. *English Studies*, 89 (3), pp.273–86.
Langbauer, L. 1999. *Novels of Everyday Life: The Series in English Fiction, 1850–1930.* London: Cornell University Press.
Marrot, H.V. 1936. *The Life and Letters of John Galsworthy*. New York: Charles Scribner's Sons.

Matz, J. 2001. *Literary Impressionism and Modernist Aesthetics*. Cambridge: Cambridge University Press.
Miller, J. 1994. *Rebel Women: Feminism, Modernism and the Edwardian Novel.* London: Virago.
Pallette, D. 1959. Young Galsworthy: the forging of a satirist. *Modern Philology*, 56 (3), pp.178–186.
Pennebaker, J. 1989. Confession, inhibition and disease. In L. Berkowitz, ed. *Advances in Experimental Social Psychology.* New York: Springer, pp.211–244.
Peters, J.G. 2001. *Conrad and Impressionism*. Cambridge: Cambridge University Press.
Saunders, M. 2004. Modernism, Impressionism and Ford Madox Ford's The Good Soldier. *Études Anglaises*, 57, pp.421–437.
Sauter, R. 1967. *Galsworthy the Man: An Intimate Portrait*. London: Peter Owen.
Shklovskij, V. [1917] 1986. Art as technique. In R. Con Davis, ed. *Contemporary Literary Criticism: Modernism Through Poststructuralism*. London: Longman Press.
Soja, E. 1996. *Thirdspace: Journeys to Los Angeles and Other Real-and-Imagined Places.* Oxford: Basil Blackwell.
White, H. 1980. The value of narrativity in the representation of reality. *Critical Inquiry*, 7 (1), pp.5–27.
White, M. and Epston, D. 1990. *Narrative Means to Therapeutic Ends*. London: Norton and Co.
Wylie, J. 2010. Writing through landscape. In C. Brace and A. Johns-Putra, eds. *Process: Landscape and Text*. Amsterdam: Rodopi, pp.45–64.

Part III

En-route writing: Writers at work

7 Trollope's *en-route* writing

Movement is integral to the idea of the scene. We gained a sense of this in Chapters 1–3 through the way in which the place of writing, as a relational space, was intimately connected to the world outside. This was through the objects that arrived and got archived within it, the material traces that littered its surfaces and the people who visited and dwelled within it, but it is also through the things, the people and the ideas that issue from it into the world; in other words, the way the study goes forth in the world. If we dispute the romantic idea of the writer in their garret, through acknowledging the things that wind up there and influence the creative process, we must also consider how the study, the garret or the studio become enmeshed, extended and transported through the mobility of its objects and its occupants. The scene of writing therefore, is constituted not only through the movements that happen within and through a place, but through those that simultaneously happen across space.

Histories of geography, and science more broadly, have long recognised how geographical knowledge is made on the move. We know how travel transforms equipment and method (Livingstone 2003), how the laboratory was replicated on scientific voyages and how important fieldwork and collecting practices were to geography's localised making (Lorimer and Spedding 2005). In these instances, 'going forth' is a recognised, normative and recordable element in the generation of knowledge. It is not however, an unproblematic process. As Charles Withers (2011) observes, in the context of travel writing, we still know relatively little about the relationship between going-forth, or being '*en-route*' as he terms it, and route writing: between what is seen and experienced and what and how this is recorded. As such, we need to understand more about the route to writing: to what writers did within the world, to their embodied experiences and to how these experiences shaped the way they wrote the world.

In part this is because one individual's *en-route* practices will never be those of another. The ways of going forth in Africa will be different to those in the Arctic, not only because of spatial and temporal differences, but because of the different gathering of people and practices that arise therein. Moreover, as each route to writing returns home and readies for publication inimitable relationships arise between author, publisher and audience (MacLaren 1992; Withers 2011). This is not, however, the only reason why we need to know more about the route to writing. Another is that *en-route* writing is not something unique to histories of geography, science or travel writing. It is found in the literary sphere as well, yet a combination of factors, from the death of the author (Barthes 1977) to fears that such work is little more than a return to early literary geography that sought verisimilitude between the real and the imagined world (Darby 1948; Gilbert 1960) may well be causal to literature's absence from such debates. Yet we know that writers wrote on the move, and in the same way that critical attention has been directed to their studies as important sites in the generation of their imagined worlds (Fuss 2004), we need to understand more about the more expansive theatres of composition they operated within. Following Withers then, *en-route* is employed here as a way of accessing the route from world to word, or put another way, as a process of mediation: in how, and through what, inscriptive practices, representational strategies and embodied experiences the world is rendered on the page.

Writing *en-route*

The relationship between Anthony Trollope's novels and his travels has been quite extensively explored. This has ranged from studies of Trollope's travel writing that examine the testimony of these works (Aguirre 2015) through analysis of how travel within his novels functions within larger discourses of British foreign and imperial policy (Chappell 2016; Goodlad 2009) to detailed study of Trollope's postal plots and the manner in which the circulation and exchange of letters created communicative uncertainty within the narrative (Thomas 2012). Interestingly though, particularly given the widespread recognition that Trollope was an unusual Victorian and writer – a prolific international traveller and one able to write almost anywhere, be it train, boat or carriage –, we know very little about his routes to writing. That is, the way his writing emerged and transformed over space. The idea of route writing, or *en-route* writing, has its origins in I.S. MacLaren's (1992) interest in the linear progression of travel writing from field notes, through a fuller journal, to draft manuscripts and a published book. More recently, Withers (2011) has called attention to the need to understand more about the practices, techniques and technologies of writing in the production of geographical knowledge, leading him to conceptualise *en-route* writing as something more than the correspondence of lived geographies with textual rendition. It is more interesting, Withers argues, to conceptualise *en-route* writing as a process of mediating the world that probes the relationship between narrative and practice and draws attention to the material practices of writing that take place abroad and at home.

Folding these debates into the literary sphere demands caution, not least because of the epistemic difference between scientific and literary conduct; unlike the scientist or explorer, the novelist is not concerned with proving the truth of the seen world or their veracity as its viewer. Instead, they are likely to be concerned with how to capture the world in tone, voice or colour so that a convincing place, plot or narrative emerges. What is more, where the field explorer is keen to discover a certain place, or prove a certain geographical theory, and so tends to record their travels in exhaustive, precise and checkable detail, the novelist is rarely driven by such an objective, placing greater emphasis on developing their own unique and inimitable voice. This does not mean that the question of mediation from world to page becomes moribund within the literary sphere, but rather that it takes on a different hue.

It is a hue clearly articulated by Fiona Stafford in *Local Attachments*, where, reflecting on Seamus Heaney's Nobel Laureate lecture *Crediting Poetry*, Stafford draws attention to Heaney's emphasis on creative work being 'equal to *and* true' (2010: 3, emphasis in original). What this means is that it offered a world that 'embodied an utterly convincing truth to life. It was rooted in something recognisable' (2010: 3). Being able to achieve this, persuasively, in poetry or prose depends, Stafford observes, on local specifics. Yet this does not mean that writing becomes insular or understandable only to those who know of the locations that are written of; rather, the power of this kind of prose lies in the way it offers a mirror to one's own world. In expressing the particular it simultaneously expresses the universal, or as Heaney eloquently put it, a text's 'trustworthiness and its travel-worthiness have to do with its local setting' (cited in Stafford 2010: 4). Within a literary sphere, then, mediating world to word is similarly bound up with ideas of how truth can travel, yet this is not so much an objective, or perhaps more aptly tangible and touchable truth, as a subjective, catalytic one; it is about conveying not a shared story, but of calling forth a shared affect or energy. Jonathan Bate (2000) terms this the difference between experiential and descriptive prose; the purpose is not to tell, but to show and reveal a world that is unique and is convincing and plausible because of this uniqueness. Put another way, the textual world may be alien, local references may be missed or dialects challenging, but these are part of the texture of the place and contribute to the sense of its realness. This type of truth is made not through having the proper equipment to hand, or measuring the world following the correct method, instead, it is about rendering or, in the language of Withers (2011) mediating, what is local in ways that resonate with people and places elsewhere. It is to these ideas of mediation that we now turn; to how and through what representational and inscriptive practices Trollope rendered his world on the page.

Trollope's travel diaries

If one aspect of Trollope's novels stands out to a geographer, it is the way they register travel and mobility. Although it has been observed that for so prolific a traveller, Trollope's novels are far from saturated with international travels (Buzard 2010), local journeys and smaller scale movements are integral to the unfolding of many of his works. A clear example of this comes in Chapter five of *Framley Parsonage*, which opens with a letter that:

> went into Barchester by the Courcy night mail-cart, which, on its road, passes through the villages of Uffley and Chaldicotes, reaching Barchester in time for the up mail-train to London. By that train, the letter was sent towards

> the metropolis as far as the junction of the Barset branch line, but there it was turned in its course, and came down again by the main line as far as Silverbridge; at which place, between six and seven in the morning, it was shouldered by the Framley footpost messenger, and in due course delivered at the Framley Parsonage exactly as Mrs. Robarts had finished reading prayers to the four servants.
>
> ([1860] 2000: 48)

Or another example from *The Bertrams*, when we follow George Bertram on his leisurely return home:

> George Bertram did not return directly to England. Since he had been in Turkey, he had made arrangement by letter with his friend Harcourt to meet him in Tyrol, and to travel home with him through Switzerland. It was about the middle of June when he left Constantinople, and Harcourt was to be in Innsbruck on the 5th August. . . . From Constantinople Bertram went up across the Balkans to the Danube, and thence through Bucharest into Transylvania . . . [h]e then again struck the Danube at Pesth; remained some little time there; again a week or so at Vienna; from thence he visited Salzburg, and exactly on the appointed day shook hands with his friend in the hall of the old 'Golden Sun' at Innsbruck.
>
> ([1859] 1991: 145–146)

What do these detailed depictions of travel, of both object and person, bring to the story? In both cases, Trollope could easily have had the letter merely appear in Framley Parsonage, likewise Bertram could just have reappeared in England; rather though, Trollope is keen to give spatial substance to his fictional worlds. His are novels not of locations but of journeys, physical and metaphorical, designed both to show how things happen not that they have happened, and how things connect not that they are connected. What we find in Trollope's novels, then, is an attempt to convey the fullness of lived time.

In part, we can trace this to the evolving nature of the nineteenth-century novel, which as Rosa Mucignat (2013) observes, was experimenting with how to heighten narrative visibility and materiality. What we find is a form that sought to widen its scope of vision and increase its spatial accuracy. This does not mean that it sought representational accuracy *per se*, although one could often enact realism's imagined journeys within the lived world, but that it attempted to offer such fulsome and exhaustive descriptions of place so that the represented world gave the appearance of completeness, concreteness and realness. The purpose was to create a spatiality that gave 'the impression of reality . . . without referring to the extra-literary world in a parasitical manner' (Mucignat 2013: 14). This only partially explains why Trollope gave us detailed journeys that were not always narratively important. Another reason, it can be argued, lay in his own practice of spatial apprehension, and to understand this, it is worth turning to Trollope's own travels and the way he registered and recorded them.

To do this, we need to begin in 1841 when Trollope began working as a Post Office Surveyor's clerk, graduating to Assistant Surveyor in 1844 and Surveyor in 1854[1]. An integral part of the surveyor's job was to travel in order to assess and improve the efficiency, reach and regularity of the postal system across all parts of the nation. The official outcome of these travels was the Postal Survey: a tabulation of the postal activity within place in terms of letters sent and received and distance travelled by the post carrier. The postal surveys were part of what Nikolas Rose and Peter Miller (1992) term, the technologies of government: the processes and procedures, geographical and bureaucratic, through which governments sought to extend and make calculable their power and influence. They were, like the railway timetable or the Ordnance Survey, part of the nineteenth-century culture of precision: the desire to know and generate a national space and time. In composing these surveys the surveyor's role was highly autonomous and proving one's professionalism and working out one's travel allowance depended upon the assiduous accounting of time, place and distance travelled. This Trollope did in his travel diaries, which he kept from 1841 to 1871 and within these we find an equally precise accounting of Trollope's daily practices over this 30-year period.

Octavo in size and so intended as something more than a pocket book, the diaries were usually leather bound denoting their need for resilience as portable and temporally durable objects. Each diary tended to cover a six-year period and so would experience much carrying between places. The tabulated layout of the diaries, with columns for the day, the date, the destination, miles travelled, coach expenses and tavern expenses, was not intended to invite lengthy prose but rather, as with the work plan, to signal information (see Chapter 5). Like the postal surveys that they underpinned (and it is interesting to observe that these had a very similar tabulated form and content), the travel diaries were modes of accounting for the self through space. Since the development of more precise forms of temporal accounting during the eighteenth century, textual forms had similarly attempted to capture the fullness of lived time. Part of the reason, Stuart Sherman (1996) argues, was to chart and explore the making of the self over time and place; it was a prompt to self-reflection. Yet it was also a way of insuring and justifying the self in a society that was recognising book-keeping as a source of social power and spatial knowledge. Histories of science stress the importance of precise accounting in the generation and verification of knowledge at this time, and it can be argued that Trollope's travel diaries emulate this epistemological practice (Livingstone 2003; Driver 2004). As a surveyor engaged in extending the postal system the travel diaries could, if necessary, be called upon to corroborate the travels that they underpinned.

What transpires within the pages of the diary is a spatial continuity. Trollope does not merely disappear from one place and reappear in another, what Sherman (1996) terms hypertravel, or we might term, dislocation (unless it is a relatively short journey); rather, the diary records movement through the world. This is not a register of movement of the type Sherman finds in Pepys' diary where multiple events are linked through a variety of connective phrases, such as 'from whence to . . .'; instead, Trollope's register of movement is more economical and made purely through choice of prepositions. He will record for instance, being 'at sea' or in transit as he does for Saturday 30 and Sunday 31 January 1858 with the words: 'To Paris' (Trollope 1858). The diaries do not, however, record movement *per se*; they record the trajectory of the journey, as indicated by the use of the preposition 'to' or dwelling-in-motion as indicated by the use of 'at', but unless Trollope was at sea or recorded his coach expenses, we get little sense of how he moved through the world. This can, of course, be intuited from other sources, for instance we know he was a frequent rail passenger and when in Ireland did much of his travel on horseback. The diaries privilege the condition of moving and not necessarily the form of this movement. What is more, the diaries do not record any detail about the place. We rarely learn anything of what Trollope did or saw whilst at sea or in Paris. There are exceptions such as when Trollope was in the Holy Lands and recorded going to the pyramids, but like his notation of place this was registered purely as a name: 'At Cairo (The Pyramids)' (Trollope 1858). It could be argued that place, as a site of activity and meaning making, is not privileged in these diaries; rather, it is place as a series of networked sites, of relationships, in which near and far places are drawn together creating, if we take the diaries as a whole, a word map. This is not a word map in the tradition of Howard Horowitz (2011) – a cartographic space given shape through words – but rather, a textual bridging and contracting of the space between, wherein what appear on maps as separate, discrete and distant places are transformed into nodes within a larger network of relationality.

Trollope's travel diaries, then, traced his trajectory, joining places up as his travels unfolded; movement on the ground was encoded on the page and, if we return to *Framley Parsonage* and *The Bertrams*, it can be argued that this in turn became encoded in the spatial imaginary of Trollope's fiction. In detailing the journey of the letter and of George Bertram, these two novels are momentarily turned into inventories of spatial knowledge. Yet it is spatial knowledge of a particular kind. The reader encounters an evanescent version of place. We learn little of the landscapes that the letter and George travelled through; places are fleetingly registered, sewn together into a panorama of significant sights, as we are moved towards the anticipated destination and arrival. Panoramas were popular representational forms in the nineteenth century (Byerly 2007; della Dora 2007). As material objects they condensed the world, bringing the different and the distant into the here and the now and presenting the unknown in a quick succession of shifting perspectives, wherein detail was dissolved into a larger, comprehensible whole. Panoramas were simultaneously disorientating and sense making devices; they confronted the reader (or observer) with a different world view, but it was a pared down world, a world without fuss or detail so that the reader was able to read themselves into the landscape with ease. Trollope's travel diaries, it can be argued, were where this representational strategy was practised; they did not transport deep or detailed knowledge of place, rather, they transported a framework of connections, continuity and opportunity. Transposing this notation of place into *Framley Parsonage* and *The Bertrams* was a way of acknowledging the contracting of the world and the bringing of the distant home, whilst appreciating that for many this world was still too overwhelming and so needed to be synthesised and presented in a graspable manner. Thus, where scientists may well have wanted to measure and record in extensive detail the geography between places, Trollope distilled and simplified the geography 'between' into something knowable. The distant was not exoticised nor its difference accentuated; instead, it was made familiar and near through the chain by which it was connected to the local. In ordering the world and performing geographical proximity, the travel diaries proffered a different way of writing distance and difference.

Trollope and the Post Office

Something of Trollope's postal activities and their relation to writing is already known, but as Withers (2011) notes, how we mediate the world is also bound up with how we experience it. Thus, we need to know something of how Trollope's postal work, beyond his practice of diary-keeping, shaped his writing. Something of this is alluded to by Coral Lansbury (1981), who, in a work that explores the influence of legal practices and protocols on Trollope's writing, suggests that Trollope's time as a post office surveyor was formative in his development of a very analytic and interrogative style of narration. This Lansbury explores primarily in relation to Trollope's travel writing, arguing that his surveyor's need to know the locality and the locality's mode of inhabitation encouraged an unromantic, realistic and relatively fair-minded depiction of other people and places. It is a relationship though that operated more widely, for arguably Trollope's postal practices were formative to his creative writing more broadly.

When he became a postal surveyor, Trollope's main role was to help improve the reach and efficiency of Britain's postal service. A key part of this was the need to know Britain's postal walks; these were the routes that postal carriers had to take in order to collect and deliver the post within their geographical area. Some areas, usually those with notable residents, had relatively regular service routes, while others had none at all, and Trollope's role was to explore how these unconnected areas could be connected. It was a role that demanded a willingness to traverse, intimately and in detail, a region and identify its postal ecology. This he began in Devonshire in 1851, after which he:

> visited, I think I may say, every nook in the county, in Cornwall, in Somersetshire, the greater part of Dorsetshire, the Channel Island, part of Oxfordshire, Wiltshire, Gloucestershire, Worcestershire, Herefordshire, Monmouthshire, and the six southern Welsh counties. In this way I had an opportunity of seeing a considerable portion of Great Britain, with a minuteness which few have enjoyed.
>
> (Trollope [1883] 1980: 88)

Unsurprisingly, postal surveyors were relatively autonomous, this was not an office job but a field job. Many surveyors were not as assiduous as Trollope, but as he advised an aspiring postal clerk: 'Do not lose any opportunity when you are at country offices of getting up the rural posts. Make a point of walking them all with the men whether you are employed to do so, or no' (Trollope 1858b). Walking the routes allowed the postal surveyor to fully understand how post moved locally and internationally, how it was conducted through space and the pressures and challenges the local landscape presented to the postal carrier, and in executing this undertaking Trollope was 'altogether in earnest' ([1883] 1980: 92). In his postal work, then, Trollope was something of a stickler for detail, requiring first- rather than second-hand knowledge of the world. What is more, he was zealous in his undertaking, keen to extend the postal routes as far as he could and 'anxious for good results'. New postal routes were not to be established where the letters carried would not cover the costs of the postal carrier, yet as 'the counting was in our own hands . . . an enterprising official might be sanguine in his figures. I think I was sanguine' (Trollope [1883] 1980: 89).

Trollope, then, was rigorous in both his data collection and exegesis; he was keen to move beyond the abstract sense of a place – as something that appeared merely as a name on a map – and understand, instead, its substantive relationships, internal logic and modes of inhabitation. We find a similar approach in Trollope's fiction, which tends to begin with the big picture, the meta-data if you like, before focusing in on its many parts. Take, for instance, *Framley Parsonage*, the fourth novel in the Barsetshire series. It was the first of Trollope's novels to be published in serial form, appearing in *The Cornhill Magazine* between January 1860 and April 1861. As his first foray into serialisation, Trollope had to reconsider his compositional practice, ensuring that each issue, which comprised three chapters, was whole both in and of itself and worked in unity with his overall concept of the novel (Hamer 1987). Hamer suggests that serialisation demanded that Trollope attend to his chapter endings, ensuring that each issue concluded with something to hold the reader's attention for the next instalment, yet it also demanded that Trollope familiarise his readers with the general contours of his narrative space. Trollope has long been associated as a novelist of character not plot (Kincaid 1977), often revealing the lineaments of his story early on, before exploring the many dimensions of these features. In *Framley Parsonage* it could be argued that this exposé or overview needed to be achieved more quickly than previously, so that on concluding the first three chapters the reader had a sense of the narrative. Thus, by the end of Chapter three, Chaldicotes, the reader is aware of the history and current situation of Mark Robarts, the novel's hero; we know of the environs of Framley and its principle residents, most notably Lady Lufton and her opponent the Duke of Omnium; and as Chapter three closes the first tension point arises, with Mark's invitation to join a party hosted by the Duke of Omnium, a man reviled by Mark's patron Lady Lufton, raising the spectre of cleavages within the close-knit community.

What is also notable about *Framley Parsonage*, at least when published as a book, was the paratextual material Trollope introduced[2]. Reflecting on the novel's composition, Trollope observed his simultaneous production of a 'map of the dear county' ([1883] 1992: 154) remarking that it had 'become necessary . . . to provide a map of Barsetshire for the due explanation of all . . . [its] localities' (2000: 167).

That this map appears in the novel's pretextual material suggests that like the opening three chapters, it is intended to give us an overarching view of Framley: we can see how Framley and its environs are organised, which parts are proximate to which, where roads and railways run and the village and county's orientation. It can be argued that Trollope's approach to this novel's beginning was akin to the surveyor's initial approach to his region. The surveyor's first encounter with his region was usually abstract, it was presented as a geographical idea, as opposed to a known place. Trollope for instance, was offered the Devonshire posting whilst still performing his surveyor's role in Ireland. There is little to indicate he had any familiarity with or knowledge of Devonshire prior to taking up his role, and it is likely that his first conception of this region was through some cartographical representation that gave an overview of the locality. Once *in situ* it was the surveyor's role to give substance to this cartographical outline, and to come to know it not as an abstract entity but as a living one. Due to the autonomy of the surveyor's job, the extent to which knowledge of a region was built up varied from surveyor to surveyor, but as Trollope made clear 'I did my business after a fashion in which no official man has worked, at least for many years' ([1883] 1992: 88). Trollope was keen to see the many nooks and crannies of this region, its many dimensions and unlike other surveyors who kept to established road systems, it was Trollope's delight to make his postal carriers go 'here and there across fields', exploring all manner of 'short cuts' ([1883] 1992: 89–90) in order to bring as much of the locality into the postal network as possible. By this method, Trollope was able to skim yards from the miles between places, and link up places in ways those who adhered to the road network might not have considered. Thus, what Trollope created was a counter-intuitive geography of his region, presenting it through a different lens so that its many dimensions became clear.

This strategy of oscillation was found in his fiction as well. If we return to *Framley Parsonage* this was a novel, P.D. Edwards (2000) observes, that even by Trollope's standards had a very flimsy plot. Trollope, as already noted, is known for his preference for characterisation over plot (Kincaid 1977), believing that the causality on which a plot depended was unrealistic. Instead of an overriding plot *Framley Parsonage* is a patchwork of sub-plots that share space and time and sometimes cross paths with one another. There is the strained relationship that unfolds between Mark Robarts and Lady Lufton, as the former behaves in a way that offends the patronage of the latter; there is Mark's sister, Lucy's romance with Lady Lufton's son; there is the tension between East and West Barset and its manifestation in the conflict between high and low church Anglicanism, and between Whig and Tory; and there is, like the Barset series more broadly, Framley's struggle to resist the power of metropolitan London. On their own, these sub-plots give the reader a multifaceted vision of Framley life, more than this though, they give us a multifaceted vision of each character, and this was Trollope's forte as a novelist:

> The skill with which he gives us view after view of his different characters, each looking, at first, as if it were the old view over again, but proving before

> long to have a something added, which gives you a sense of a completer knowledge of the character, reminds us of nothing so much as the zigzags of a road terraced up a steep hill-side from which you are constantly getting the same view of a valley repeated again and again, but each time with some novelty of aspects and additional command of its relation to other neighbouring valleys.
>
> (Anon. 1867: 18)

The analogy this review draws between the unfolding of the narrative and the unfolding of a landscape vista returns us, in a metaphorical sense at least, to the idea of *en-route* writing, for we find in Trollope's form a method akin to that which he exercised on his postal walks. His finding of short cuts, his preference for going off the beaten track, and his delight in coming upon unsuspecting postmasters at odd times of the day and from odd directions disconcerting them into giving candid information on the state of their local postal network (Trollope [1883] 1992) enable us to draw parallels between how Trollope moved within and knew the world and how this shaped, in profound but subtle ways, the form and development of his fictional narratives.

What should also be noted about these sub-plots is the way in which they intertwine, for as Skilton (1972) argues, Trollope's fiction works through the interaction between characters. Rather than developing his characters as individuals he develops them through the shifting perspectives of the consciousness of other characters. This can be seen quite clearly through the unfolding characterisation of Framley Parsonage's hero, Mark Robarts. One of our earliest perspectives of Mark comes from his wife, Fanny, who tries to counsel him not to accept an invitation to a gathering at Chaldicotes, home of Mr Sowerby and centre of Whig politics, for fear of offending Mark's patron, Lady Lufton:

> 'Just now . . . I answered a letter that was brought to me from Chalicotes. Sowerby insists on me going over there for a week or so.'
>
> 'Go to Chaldicotes for a week, Mark? . . . It is just what Lady Lufton won't like.'
>
> (Trollope 2000: 9)

Mark rationalises his acceptance of the invitation as it will afford him an opportunity to meet the bishop. Yet Mark's desire for professional advancement, his natural desire to know his professional seniors and his awareness that, as a man of God, he serves not just his immediate parishioners but the wider Framley community, is countered later in novel, when Lady Lufton learns of Mark's deception:

> 'Oh he is going to the palace, is he? Well; he must choose his own friends. It's a pity he did not see Miss Proudie before he met you, he might have lived to be the bishop's chaplin. Gatherum Castle! You don't mean to tell me he is going there? Then I tell you fairly, Fanny, that I have done with him.'
>
> (Trollope 2000: 53)

What to Mark was professionalism and, veiled to a certain degree as Christian integrity, was regarded in Lady Lufton's eyes as religious betrayal. As the novel progresses further dimensions to Mark's character are made apparent. Befriended by the Chaldicote set, he is offered a prebendal stall in Barsetshire, which as he informs his wife 'don't generally go begging long among parish clergymen. How could I reconcile it to the duty I owe my children to refuse such an increase to my income?' (Trollope 2000: 213). As *Framley Parsonage* progresses, Robarts' professionalism and Christian virtue begin to appear, at least to some, as a little besmeared by Mammon. The relationality inherent in Skilton's argument about character resonates with Hones' argument about the making of narrative space, which 'has more to do with the ways in which lives and events interrelate than it does with specific fixed settings; its world is clearly a world of coexistence; and the world in which the events take place is clearly in a state of becoming' (2014: 80). What this means, as Hones goes on to observe, is that it allows us to see how characters connect in unexpected ways, it draws together, in equally unforeseen ways, spaces that seem far apart and it adds a depth to space, as relationships hint at others that are happening off-stage. As with Trollope's postal walks that resist an easy journey from A to B and instead seek out novel connections that generate new ways of knowing place and of bringing marginal places into new relational constellations, his relational development of character creates a method of slow disclosure that structurally resists any sense of a linear narrative trajectory or a discrete narrative space. Thus, if Trollope's pre-text and early narrative suggest order, both social and spatial, as *Framley Parsonage* develops this order is challenged, upturned and rearranged as the reader is invited to enter or view the narrative space from different perspectives: we come to know it through its parts not its whole. As Mucignat (2013) observed, nineteenth-century realism sought to heighten visibility and offer a spatiality that gave the appearance of reality. This could be done through description and the detailed rendering of place, but *Framley Parsonage* suggests it could also be achieved structurally, with Trollope's spatial and temporal structure being driven not so much by a 'generic' seen world as by his epistemological becoming within it.

Conclusions

'Once there is thought to be some specially intimate relationship between the world of a Trollope novel and the real world, it follows that the work of art itself disappears from view, becoming, as it were, totally transparent' (Skilton 1972: 138). To consider Trollope's work in relation to the idea of *en-route* writing might seem tantamount to eclipsing the text in preference for the life of the author, but as work in histories of science and geography demonstrate, the more we trouble, rather than reify, the boundary between the real and the imagined world the more other dimensions of the work of art come into view. From the perspective of *en-route* writing these other dimensions are often epistemological rather than ontological; they let us explore not what texts are but how they come to know and articulate the world.

The relationship between knowledge and literature is a problematic one, not least because knowledge is often taken as a coherent, objective entity and literature an imagined world in which knowledge, as something real and verifiable, has no place. Yet, in a provocative series of essays, Michael Wood (2005) posed the question, what does literature know? In his various attempts to answer this question, Wood argues that like the sciences, literature has its own form of hard knowledge: it tells us truths about the world. Often these are the 'other side': the other case, the things we may not want to hear or know of, or, to put it another way, it often gives us troublesome knowledge. Although Wood is less concerned with how this knowledge arrives within the text, focusing on knowledge as something pre-existing within the text, or made through the coming together of text and reader, his work is instructive for the way it destabilises the assumption that literature is only and exclusively the world of the imagination; epistemologically the novel spills beyond its boundaries. If literature knows things, it seems natural to ask, how does it know? Posing this question draws us to the author, but it equally draws us to issue of technique, inscription and form. Bringing these two together through the lens of *en-route* writing allows us to see technique, inscription and form not as pre-existing facets of a text, but as parts of its making.

As Withers (2011) observed, there is still much to know about the relationship between being *en-route* and route writing. This lacuna is clearly evident in relation to literary texts, where disciplinary heritage and qualitative differences in writers' relationships to the world make the idea of *en-route* writing a nebulous one to work with. A route through the world is not, for instance, a route to writing about it; rather, it is more likely that aspects of the route may be worthy of notation, they might impress upon the senses, or, and as was the case with Trollope, shape the text in structural and formal ways rather than mimetic ones. By the very nature of their craft, imaginative writers have no will to present their world as truth, but this should not discount that a relationship between the real and the imagined exists, that one informs the other, and that literature knows something of this real world. More importantly, though, this world has travelled into the imagined world in ways that tell us something about both the practice of creative writing and the ways in which spatial knowledge is encoded, at multiple scales, within the imagined world. Knowing something of the epistemic basis of the imagined world is not about reducing it to biography, but rather, about posing new questions about representational techniques and enriching our understanding of the creative process.

Notes

1. Trollope began working for the Post Office in 1834. In the latter part of this decade the Post Office underwent a series of reforms in light of Rowland Hill's 'Post Office Reform: Its Importance and Practicality' (1837). The most notable of these was the introduction of the Penny Post.
2. The map Trollope produced was reproduced by one of his earliest biographers, Michael Sadleir in *Trollope, a Commentary* and has subsequently appeared in various editions of *Framley Parsonage*, most notably the World's Classics editions.

References

Aguirre, R. 2015. 'Affairs of state': mobilities, communication, and race in Trollope's The West Indies and the Spanish Main. *Nineteenth Century Contexts*, 37 (1), pp.1–20.

Anon. 1867. Books. *The Spectator*, (21 September 1867), pp.18.

Bate, J. 2000. *The Song of the Earth*. London: Picador.

Barthes, R. 1977. *Image, Music and Text*. New York: Hill and Wang. Trans. S. Heath.

Buzard, J. 2010. Portable boundaries: Trollope, race and travel. *Nineteenth Century Contexts*, 32 (1), pp.5–18.

Byerly, A. 2007. 'A prodigious map beneath his feet': virtual travel and the panoramic perspective. *Nineteenth Century Contexts*, 29 (2–3), pp.151–168.

Chappell, L. 2016. Anthony Trollope's narrative temporalities and the emergence of the Middle East. *Literary Interpretation Theory*, 27 (1), pp.29–49.

Darby, H.C. 1948. The regional geography of Thomas Hardy's Wessex. *The Geographical Review*, 38 (3), pp. 426–443.

della Dora, V. 2007. Putting the world into a box: a geography of nineteenth-century 'travelling landscapes'. *Geografiska Annaler. Series B, Human Geography*, 89 (4), pp.287–306.

Driver, F. 2004. Distance and disturbance: travel, exploration and knowledge in the nineteenth century. *Transactions of the Royal Historical Society*, 14, pp.73–92.

Edwards, P.D. 2000. Introduction. In A. Trollope, *Framley Parsonage*. Oxford: Oxford University Press.

Fuss, D. 2004. *The Sense of an Interior: Four Rooms and the Writers that Shaped them*. London: Routledge.

Gilbert, E.W. 1960. The idea of the region: Herbertson Memorial lecture. *Geography*, 45, pp. 157–175.

Goodlad, L. 2009. Trollopian 'Foreign Policy': rootedness and cosmopolitanism in the mid-Victorian global imaginary. *Modern Language Association*, 124 (2), pp.437–454.

Hamer, M. 1987. *Writing by Numbers: Trollope's Serial Fiction*. Cambridge: Cambridge University Press.

Hones, S. 2014. *Literary Geographies*. London: Palgrave Macmillan.

Horowitz, H. 2011. Wordmaps. In M. Dear, J. Ketchum, S. Luria and D. Richardson, eds. *GeoHumanities: Art, History, Text at the Edge of Place*. London: Routledge, pp.107–111.

Kincaid, J.R. 1977. *The Novels of Anthony Trollope*. Oxford: Clarendon Press.

Lansbury, C. 1981. *The Reasonable Man: Trollope's Legal Fiction*. Guildford: Princeton University Press.

Livingstone, D. 2003. *Putting Science in its Place: Geographies of Scientific Knowledge*. London: University of Chicago Press.

Lorimer, H. and Spedding, N. 2005. Locating field science: a geographical family expedition to Glen Roy, Scotland. *British Journal of the History of Science*, 38 (1), pp. 13–33.

MacLaren, I.S. 1992. Exploration/travel literature and the evolution of the author. *International Journal of Canadian Studies*, 5 (1), pp.39–68.

Mucignat, R. 2013. *Realism and Space in the Novel, 1795–1869*. Farnham: Ashgate.

Rose, N. and Miller, P. 1992. Political power beyond the state: problematics of government. *The British Journal of Sociology*, 43 (2), pp.173–205.

Sherman, S. 1996. *Telling Time: Clock, Diaries and English Diurnal Form, 1660–1785*. London: Chicago University Press.

Skilton, D. 1972. *Anthony Trollope and his Contemporaries*. London: Longman.

Stafford, F. 2010. Local attachments: the province of poetry. Oxford: Oxford University Press.

Thomas, K. 2012. *Postal Pleasures: Sex, Scandal and Victorian Letters*. Oxford: Oxford University Press.

Trollope, A. [1883] 1980. *An Autobiography.* M. Sadleir and F. Page, eds. Oxford: Oxford University Press.

Trollope, A. [1883] 1992. *An Autobiography*. Oxford: Oxford University Press.

Trollope, A. [1860] 2000. *Framley Parsonage*. Oxford: Oxford University Press.

Trollope, A. [1859] 1991. *The Bertrams*. G. Harvey, ed. Oxford: Oxford University Press.

Trollope, A. 1858. *Travel diary, Vol.4: 1855–1861.* M. L. Parrish Collection of Victorian Novelists (C0171), Manuscripts Division, Department of Rare Books and Special Collections, Princeton University Library.

Trollope, A. 1858b. Letter from Trollope to B. Blake Esq (1 October 1858). *Letters; circa 1839–1882; M. L. Parrish Collection of Victorian Novelists, Volume 1, 2 & 3*. Princeton: Manuscripts Division, Department of Rare Books and Special Collections, Princeton University Library.

Withers, C. 2011. Travel, en-route writing, and the problem of correspondence. In M. Hvattum, J.K. Larsen, B. Brenna and B. Elvebakk, eds. *Routes, Roads and Landscapes*. Farnham: Ashgate, pp.85–98.

Wood, M. 2005. *Literature and the Taste of Knowledge*. Cambridge: Cambridge University Press.

8 Galsworthy's epistolary practices*

The relational making of *Fraternity* (1909)

In his study of the writing practices that lay behind George Chalmers' Caledonia, a historical and geographical study of Scotland, the first volume of which was published in 1807, Charles Withers (2004) draws attention to the writing *in* as opposed to the writing *of* the text: to the way in which Chalmers relied upon the geographical knowledge of others, conveyed in letters, to compose his text. It is a work that builds upon an interest in the eighteenth-century Republic of Letters, a community of scholars who used the epistolary form to circulate and develop scientific knowledge in mutual and reliable ways. Withers' study of Caledonia, however, brings a greater geographical cognisance to these debates. It suggests that where letters are written, the conventions this may place upon the form and the manner in which they move all work to shape geography in particular ways. In considering the place of geography in the Republic of Letters, Withers poses the questions of how 'different subject areas work through letter writing in terms of their associated social and intellectual practices?' (2004: 35). We know, from the epistolary form of many eighteenth-century novels, the importance of letters and letter writing to imaginative literature (Altman 1982). Yet letters mattered to the nature of the novel in other ways too. They mattered, as they did with Caledonia, to the very making of the novel as an imagined, material entity.

The epistolary relationships at the heart of this chapter were those maintained between John Galsworthy and Edward Garnett and between Galsworthy and Joseph Conrad. In 1908 when we first intrude upon the correspondence of these three men, they had been literary associates and friends for some years. Galsworthy at this time was engaged on his novel that would become *Fraternity* (1909). He already had five full length novels and several volumes of short stories and criticism to his name, and had carved out a particular style for himself as a new and challenging writer (Anon. 1906) who displayed 'a remarkable power of ironic insight combined with an extremely keen and faithful eye for all the phenomena on the surface of life' (Conrad 1906: 449–450). Against this backdrop, *Fraternity* was 'a more ambitious piece of writing' (Dupré 1976: 161). It was a purposeful emulation of French and Russian naturalism, a literary style that aspired to a grittier, darker and more introspective rendering of the world than that offered by Victorian social

*An earlier version of these ideas was published in The spatial event of writing: John Galsworthy and the creation of Fraternity. *Cultural Geographies*, 20 (3), pp.285–298.

realism. It was also a novel that more than any of Galsworthy's previous works was a study of character. In consequence, it was a piece of work that Galsworthy was very uncertain about, and thus, sought advice and guidance from his close friends Edward Garnett and Joseph Conrad asking them to help him 'cut through stereotyped or irrelevant language and ideas so that the real writer can be heard' (LeFevre 1987: 68). The correspondence between these three, which grew up in the world before the text, tell us something of how the fictional world of *Fraternity* was a relational making, and exploring these epistolary relationships reveals, following Hones (2008), something of the spatial event of the novel's writing.

Epistolary practice and textual composition

Authorial epistolarium have received significant attention within literary studies. Emily Dickinson's extant correspondence has been enlisted both to shed light on her compositional process and to trouble the boundary between her poetry writing and her letter writing (Salska 1998; Socarides 2012). Olive Schreiner's letters have lent themselves to studies of epistolary style and identity formation (Stanley 2004) but also, as with Dickinson, to the porosity of the boundary between letters and literary forms (Stanley et al. 2012). Other studies have looked to how letters create and sustain literary friendships (Clay 2006) and to how they are used to disseminate literary work (See Socarides 2012, but also Bushell 2009). All these studies recognise, to varying extents, the latent creativity of the letter; its capacity to hold within itself something, be it an idea, a composition or a social agency, which on reception does something in the world. This something can be self-presentation or publicity. It can be sympathy or empathy. It can be formal or stylistic development. What has accrued less interest however, is the creative potential, or, after Barton and Hall (2000), the creative consequences of letter writing, within the compositional process. Many writers wrote to friends about their writing. Sometimes this was to do little more than complain about writer's block or to note the start or conclusion of a piece of work. At other times though, writers wrote to request assistance, advice or feedback on compositions to date. This could be a search for affirmation or endorsement that was given and received quickly and relatively perfunctorily, yet it could also be a productive process that was conducted across a series of letters and in which the content of these letters came to play a critical role in the making of the material text and its narrative world.

Something of this is captured in Susan Whyman's (2007) study of Samuel Richardson's epistolary practice and the creative consequences of letter writing downstream of the eighteenth-century epistolary novel. Drawing on the idea of epistolary literacy – the practices of letter writing, reading, interpretation and response – Whyman argues that Richardson invited readers to engage with his text and suggest narrative developments. Though he rarely incorporated their suggestions *per se*, he did revise his narratives in response to their moral anxieties. One of his most significant readers and correspondents was Lady Bradshaigh whose interest in *Clarissa* led Richardson, in subsequent editions, to incorporate editorial apparatus to clarify his intentions and meaning (Kauffman 1986). The reader letters that Richardson encouraged brought to light aporias in his story and his narrative and while he did not respond to them straightaway he did deploy subsequent paratextual material to resolve the queries raised by his readers. Here though, Richardson's text was in the world; his readers were not responding to drafts but to a published text. As such, their capacity to influence the narrative world of the novel was fairly minimal. Another example of reader involvement in the compositional process comes in the serial fiction of Charles Dickens. This way of writing created a temporal and spatial juncture for others to correspond with the author and influence the turn of narrative events (Bradbury 2001). Yet as Jennifer Hayward (1997) remarks, few of Dickens' letters survive making the creative consequences of letter writing more anecdote than attestable happening. Yet in both these examples there is a hint of the creative potential of epistolary practice in the compositional process. It is though, a practice that needs further elucidation, not only to realise the relational nature of literary creation, but to understand, more fully, the productive effects or consequences of this sociality on the space of the narrative: on what gets made, on how it unfolds and on the consequences this space is intended to have within the world.

To explore the role of correspondence in the writing of *Fraternity* it is helpful to turn first to the work of Steven Shapin (1994) who, in the context of seventeenth-century science, argues that science's making owed to epistemological decorum: to the practical management of factual testimony. Science was often conducted at a distance, through field studies and explorations, and in arriving 'home' its claims to truth needed to be verified. Shapin develops seven maxims that informed this verification: 'i. assent to testimony which is plausible; ii. which is multiple; iii. which is consistent; iv. which is immediate; v. which is from knowledgeable or skilled sources; vi. which is given in a manner which inspires a just confidence; vii. which is from sources of acknowledged integrity and disinterestedness' (1994: 212). As has been alluded to elsewhere, the making of science and the making of imaginative literature have different relationships to truth (Saunders 2010). Where the former regards being-in place and seeing-for-one's-self as bound up with the production of truth and veracity, the latter positions the relationship between the real and the imagined as more one of authenticity than verisimilitude. This does not, however, preclude the need for affirmation in a vein similar to that Shapin outlines. In writing to Garnett and Conrad, Galsworthy was inviting them to evaluate the plausibility of his fictional world, to assess the consistency between parts and whole, to measure its stylistic and formal innovativeness and to judge its convincing and authentic manner. Similarly, in seeking the views of Garnett and Conrad, Galsworthy was not choosing randomly. These two men were significant figures in the Edwardian literary world and between them their professional integrity and expertise encompassed both the scene of reception and the scene of production. Thus, it is through these lenses, inspired by but slightly distorting Shapin's earlier maxims, that we begin to explore the letters that were exchanged about *Fraternity*.

Fraternity: writing a different world

The working title of *Fraternity* was *Shadows* and Galsworthy's inspiration or seed incident (Doyle 1998) was the symbiosis of London's two social worlds: that of the rich and that of the poor. Social injustice had long actuated Galsworthy and, in particular, the indifference of his own social class to the social ills quite literally on their doorstep (Dupré 1976). The novel tells the story of the Dallison family – Hilary Dallison, his wife Bianca, his brother Stephen and Stephen's wife Cecilia – as they come to realise that their comfortable, middle class lifestyle is predicated upon the existence and toil of their working-class shadows. At first glance this novel may seem part of Galsworthy's 'indubitably English' style, for as Edward Garnett observes, 'power was given him to analyse and set down the impulses and actions, mental, moral and emotional, of the English types he knew so well, as explicitly as the two English masters, Thackeray and Trollope, who drew each the Englishmen of their day, in portraits no less faithful to life' (1934: 6–7). In drawing comparisons with Thackeray and Trollope Garnett signals the stylistic continuity of Galsworthy's prose; he was a writer in the tradition of Victorian narrative realism. This created a social world, Mucignat (2013) argues, which was notable for its heightened visibility, depth and mobility; it was rendered in increasing detail, it had a verticality both socially and temporally and it was composed of discontinuous and heterogeneous spaces that fostered manifold displacements. It is an alignment though that must be read in relation to Galsworthy's later work, for it was a posthumous characterisation. As several of his biographers remark, after 1914 Galsworthy's novels lost something of their stylistic freshness and originality and this they attribute to his return to the social elite; with his divorce forgotten and society's acceptance of Ada his writing lost its motive force (Barker 1963; Gindin 1987). Thus, the novels that came after this were more conventional; more prone to creating a sweeping social world.

That said, even in relation to his pre-1914 novels the composition of *Fraternity* was, according to Galsworthy, somewhat different, observing to Edward Garnett that it was 'a queer book. More intimate than anything I've done – I mean by that, that there seems less machinery of story, less history, more life' (Garnett 1934: 164). Galsworthy came to regard *Fraternity* as one of his most closely observed pieces of work, believing it to be, 'perhaps, the deepest of them all' (Gindin 1987: 257).Within his oeuvre though, it is a novel that has attracted little attention, perhaps because, as one of his biographers remarks, 'in so far as the book is the story of character Fraternity can be seen as a modern novel in a sense that his other books were not' (Dupré 1976: 161). It is an anomaly, a novel it is difficult to reconcile with Galsworthy's oeuvre both pre- and post-1914. Although difference is the synthesiser's nightmare, and something which in Galsworthy's case it has been easier to side-step rather than engage with, as Rob Kitchin and James Kneale (2001) have observed, working with new literary styles is liberating and creative. It defamiliarises and makes strange the known, extending a writer's cognitive space and enabling them to think about new ways of writing the world. The suggested modernism of *Fraternity* for example, hints at an imagined world that is introspective, fragmented and ambiguous in nature. Moreover, it draws the novel into a very distinctive textual geography, one that is more European, transnational and mobile in nature than that associated with Victorian social realism (Hegglund 2012). Consequently, it can be suggested that the difference and unfamiliarity of *Fraternity's* style may well have afforded Galsworthy opportunities to rethink and rearticulate the representational nature of his novels. Thus, it may well be that *Fraternity* has defied easy classification because it writes the world in a way that challenges not only the indubitably English style of Galsworthy's writing, but also its indubitable Englishness.

Edward Garnett's modernist provocation

Shapin's first and fourth maxim concern the plausibility and the immediacy of observation; greater veracity is seen to lie in self-observation rather than in retelling and recycling the observations of others. Within the literary sphere plausibility and immediacy might be better framed as timeliness or relevance particularly in relation to the formal development of the novel. Edward Garnett was keenly aware of the manner in which the artistic field was developing in the early years of the twentieth century. Garnett was something of a self-made 'man of letters', having begun his career as a reader for several London publishing houses, before moving on to become an editor, critic and writer of some repute. In 1906, when his play *The Breaking Point* was refused a license, he initiated a vitriolic campaign to free art from censorship and initiate a new type of theatre that evoked the challenging style of continental playwrights like Henrik Ibsen and Gerhart Hauptmann (Gindin 1987). It was an outspokenness that may well have had its spark in Garnett's association with the international socialist movement. In 1889 Garnett had married Constance Black, a socialist and Russian translator and, in 1895, the pair moved to Limpsfield Chart, an area of the Surrey weald that was home to a growing number of socialist thinkers. Here, the Garnetts built their own house, The Cearne, which would become a popular gathering place for Edwardian literati. It was from here that Garnett helped nurture the careers of some of the early modernist writers, such as D.H. Lawrence and Joseph Conrad. Conrad, for instance, first met Ford Madox Ford at the Garnetts' and the pair went on to work together not least on *The English Review* (Conrad 1990)[1]. Moreover, it was from here that Garnett corresponded, at length, with Conrad over the need to rethink realism's modes of depiction.

Garnett and Galsworthy first met in 1898 when the latter submitted his manuscript of *Jocelyn* to Fisher Unwin. Garnett was the reader and his report on *Jocelyn* marked the beginning of a long and critical friendship between the two. In 1900 though, and perhaps in light of Garnett's conversations and correspondence elsewhere, he began to encourage Galsworthy, in the latter's subsequent story, *A Man of Devon*, to move towards a style of 'intense individuality' (Garnett 1934: 19). This was a style he continued to encourage Galsworthy to adopt, criticising early drafts of *The Man of Property* for being wedded to too explanatory a method of writing: to telling rather than showing (Garnett 1904). Bearing in mind Garnett's posthumous assessment of Galsworthy's style it is interesting that in the early years of the twentieth century he evidently believed that Galsworthy was capable of writing in a new and different way. Galsworthy's tendency to explanation though, was something Garnett flagged again in his early reading of *Fraternity*, particularly in relation to the characters of Martin and Mr Stone, who, he protested to Galsworthy, veered towards propaganda and thereby over-determined the interpretative space of the novel (Garnett 1934). Martin is a trainee doctor, committed to the idea that poverty can be obliterated by living in the slums and setting good examples of healthfulness. Mr Stone, Hilary's father-in law, is in contrast a theoretician. It is his belief that his book on universal brotherhood will demonstrate where society is going wrong. Galsworthy is very explicit in the connections he draws between these two characters:

> Martin says an enthusiast is worse than useless; people, he says, can't afford to dabble in ideas or dreams. He calls grandfather's ideas Palaeolithic . . . Martin's so cocksure. I don't think he'd find many men of eighty who'd bathe in the Serpentine all the year round, and do his own room, cook his own food, and live on about ninety pounds a year out of his pension . . . and give all the rest away. Martin says that's unsound, and the 'Book of Universal Brotherhood' rot.
>
> ([1909] 1995: 76).

In arguing that Martin and Mr Stone risk turning *Fraternity* into propaganda, Garnett is intimating that their characterisation is over-written; they are too revelatory and sit, perhaps, somewhat uncomfortably with the more self-conscious, subjective and ambiguous construction of Hilary. In telling rather than showing their contrary positions, these two characters compromise the plausibility of the literary world that Galsworthy was creating; a world that was moving away from the securities of narrative realism with its knowable world, its stable social order and its clear character motivations (Hegglund 2012). In drawing attention to the inconsistencies of Martin and Mr Stone vis-à-vis the novel's other characters, Garnett had identified a disorientating gap in the text. As Wolfgang Iser (1993) observes, gaps in the text are spaces of indeterminacy where meaning can be made. They are a very necessary part of the novel's apparatus and allow for its actualisation within the world of the reader. Gaps can be halts in the narrative or changes in view point that plunge the reader into the text requiring them to take responsibility for interpretation, for meaning is not intended to be self-explanatory. Such gaps, however, need to be stylistically consistent, otherwise an author risks bewildering their audience by presenting them with possibilities for any number of narrative worlds. Garnett therefore, is urging Galsworthy to be more consistent in his stylistic innovation in order to prevent reader disorientation.

It was something that Galsworthy initially resisted and the nature of this resistance repays closer attention, for it tells us much about the interpellation of worlds as concerns and ideas from one were transported to another. On 26 September 1908 Galsworthy wrote to Garnett observing that, 'As to Mr Stone, I'm not at present shaken. He has not given anyone but you any feeling of unreality'. In defending Mr Stone Galsworthy's use of the phrase 'anyone but you' marks Garnett's reading out as different; as exceptional within his carefully chosen audience. It is an exceptionalism that Galsworthy elaborates later in the same letter when he accuses Garnett of being a purist and 'for one purist one writes for 10,000 non-puritans (I speak in the literary sense)' (Garnett 1934: 167). By employing the common reader in his defence, Galsworthy suggests that Garnett's aesthetic ideas are out of touch; they are the ideas of a highly esteemed and aesthetically progressive figure within the London literary world and this is not necessarily the best way of reaching the larger audience that lies beyond this cloistered circle. Garnett's reply has not been traced, but Galsworthy wrote again to Garnett several weeks later to communicate his change of heart: 'I have succeeded in eliminating most if not all of what you criticized . . . it was especially hard for me in this book to see your point of 'danger', because I am not seriously after propaganda' (Garnett 1934: 168). What made Galsworthy change his mind, and how exactly he altered Martin and Mr Stone, we do not know, but what this vignette reveals is that the medium that brought these worlds together made for uncertainty: ideas did not travel unhindered or easily through the epistolary world.

Letters function to bridge gaps – spatial, temporal and imagined – that exist between worlds. They were (and are) substitutes for conversation; linking 'I' with 'you', 'self' with 'other', writer with recipient. This is what Janet Altman (1982) terms the epistolary pact; letters are written with an intended reader in mind and they call upon that reader to respond. Yet unlike conversation, letters are premised on absence, distance and suspension; the other voice in the conversation is not present, they are elsewhere in time and space and so the letter, as it travels between, suspends and extends the dialogue. Thus, where conversation facilitates and depends upon the immediate checking of understanding, the elaboration of meaning, interruptions and the seeking of clarity, there is greater ambiguity within the epistolary pact; meaning can go awry and depends upon time for explanation. Thus, in the same way that scientific ideas transform over space, what we find in the Garnett–Galsworthy correspondence is a similar transformation but of narrative worlds as interpretation deforms intended meaning. In so doing though, we must not lose sight of the fact that uncertainty also gives rise to creative possibilities. Where Chalmers issued instructions to his correspondents for the gathering and returning of information in order to lessen ambiguity, the letters between Garnett and Galsworthy were far less clear-cut, as Galsworthy's uncertainty over Garnett's point of danger indicates. Yet, as John Shotter (1993) argues, these points of uncertainty demand imaginative completion. This is not a case of just filling-in, improvising, or as Stanley et al. (2012) observe, sending a letter to ask for clarification, but rather, of exploring what they could possibly mean and this requires us to know something of their point of origin and how this point fits into the larger whole. What this meant for Galsworthy was that his comprehension of *Fraternity's* danger required him to know something of Garnett's professional and aesthetic world and to fold Garnett's letter, as a bearer of traces of this world, back into this world in order to appreciate its possible meaning. Letters, therefore, are not one-way bridges that transport ideas and then leave them; instead, they function through reciprocity not response *per se*. The bridge requires imaginative re-crossing (mental explorations) as well as textual re-crossings (in the form of written responses) in order to construct a plausible and reliable narrative world.

Joseph Conrad: the resistant reader

Garnett was not the only recipient of an early draft of *Fraternity*, so too was Joseph Conrad, who engaged in a somewhat different, and perhaps more resistant reading. Conrad's concerns align with Shapin's third and sixth maxim, which relate to the consistency of and manner in which testimony is given, and which, in a literary sphere transpose as questions of stylistic consistency and fidelity of voice. Conrad, as we know, was engaged in rethinking realism's modes of depiction, and as *The Spectator* observed of his first novel *Almayer's Folly*, it 'is a decidedly powerful story of an uncommon type, and breaks fresh ground in fiction' (Anon. 1895: 530). Unlike realism with its ordered narration that presents the world of a universal observer, Conrad was experimenting with an impressionistic mode of address that immersed the reader in the immediacy and subjectivity of experience, enabling them to encounter the world as a present happening not a reported event (Peters 2001). Given Conrad's development of an impressionistic style it is interesting that he did not endorse the 'intense individuality' of *Fraternity*. The reason for Conrad's ambivalence may lie in his perception of the novel's inconsistency; it was offering a narrative world that was too different to anything else within Galsworthy's oeuvre. As he wrote to Galsworthy, there is 'the public to consider. A writer like you must make himself accessible for the sake of that which he is working for' (Karl and Davies 1991: 127). Your style is that of 'fidelity to the surface of life – the surface of events – to the surface of things and ideas' (Marrot 1936: 114). Authorial success is dependent, Michael Curry (1996) argues, upon adherence to an accepted style, which once accreted is difficult to challenge, and in Conrad's view, the world that *Fraternity* created had little fidelity to the expectations of Galsworthy's normal audience. It was stylistically looser, more fluid and more speculative than his earlier works and this, Conrad argued, was too confusing, and if he found it confusing so too would other readers:

> [Fraternity] is anything but simple. Its suggestions are multiple . . . and run in and out of each other. Yes! It is a complicated book. It is possible to view it in innumerable ways . . . But before all it is the book of a moralist . . . [and a] moralist must present us with a gospel – he must give counsel, not to our reason or sentiment, but to our very soul.
>
> (Marrot 1936: 229–30)

The reason for urging Galsworthy towards greater conservatism may well lie in Conrad's own experiences of London literary society, for it was a society he effectively wrote his way in to. Conrad was an outsider, a Polish immigrant and a naval shipman, and on both counts this made him something of an oddity within London literary society, which was still a relatively closed social world in the early twentieth century. Conrad received British citizenship in the late 1880s and from that time he gradually began to relinquish his seafaring career in preference for a life of letters, publishing his first novel *Almayer's Folly* in 1895. *Heart of Darkness*, *Lord Jim* and *Nostromo* quickly followed, and by 1908 he had published thirteen works of fiction. Today these are considered masterpieces of the early modernist movement, yet despite critical acclaim at the time, popular and financial success as a novelist eluded Conrad until much later in his career. In order to support his writing Conrad often sought financial assistance from Galsworthy, and so, his measured realism seems unsurprising when he advises Galsworthy to think not only of art, but of reception (Gindin 1987).

The narrative world of *Fraternity* was not, therefore, stable; its meaning did not travel uninterrupted from reader to reader, location to location, rather, it was constantly remade in light of the reader's world and the way this intersected with their encounters with an author's textual heritage – the other narrative worlds they have created. Conrad, for instance, saw Galsworthy's previous novels as having a 'fidelity to the surface of life', and it was this approach that produced the kind of narrative world an audience expected to find in the novels of John Galsworthy. Take for example, *The Man of Property* (1906). This was narrated primarily by Soames and Old Jolyon Forsyte. There was some use of free-indirect style, but reported speech and a strong authorial voice that both described the scene and linked together different viewpoints was the dominate narrative style. A sense of this was evident in Soames Forsyte's discussion with his architect Bosinney, over his planned suburban retreat, Robin Hill:

> Soames hated sunshine, and he at once got up to draw the blind. Then he took his own cup of tea from his wife, and said, more coldly than he had intended:
>
> 'Can't you see your way to do it for eight thousand after all? There must be a lot of little things you could alter'. Bosinney drank off his own cup of tea at a gulp, put down his cup, and answered:
>
> 'Not one!'
>
> Soames saw that his suggestion had touched some unintelligible point of personal vanity.
>
> (Galsworthy [1906] 2001: 99)

Fraternity lacked dominant characters to equal those of Soames and Old Jolyon Forsyte or the stability and straightforwardness that these characters imparted. Perspectives shift in *The Man of Property* but these perspectives were often those of Soames or Old Jolyon, or were referred back to these two characters. In contrast, *Fraternity* gave us Hilary, who Conrad believed was not:

> a sufficiently big and human figure to stand in the forefront of the great question, the enormous interrogation point which for me symbolises the book . . . if the thesis of the book is as I understand it, you should have presented to us a man really belonging to the class and with the noble instinct trying to assert his manhood against the heart-withering, brain-muddling convention.
>
> (Marrot 1936: 233–234)

Conrad, by interpreting the novel in this way, was something of a resistant reader; he was looking for a closed narrative world, one that had a linear plot and unambiguous morality, in which he needed to do little interpretative work as a reader. This owes, in part, to Conrad's own literary practice, for as Clive Barnett (1996) argues, he tended to create narratives that were premised upon a closed community of readers, thereby allowing him to control interpretation. In *Fraternity*, Hilary subverts this possibility, for in his introspection and mobility he is insufficiently decisive and therefore unable to direct the reader or signify their construal of the narrative world. At the conclusion of the novel, Conrad expects Hilary to indict his class for their shallowness and prejudice, but rather, he returns to his class, undermining the strong narrative style and decisive morality Galsworthy establishes in his earlier novels. What we see, therefore, is not only geographical worlds, but imagined worlds impinging upon, and resisting, the creation of *Fraternity's* narrative world.

Fraternity en-route

Hones has argued that a key feature of narrative space is coexistence or, put another way, 'an understanding of the world as a set of elastic networks, distances, and relations' (2014: 70). From this perspective, the narrative world is not just a container space – a fixed, static setting in which events happen – but something that is made. In an echo of Marcus Doel, this is space, Hones suggests, 'as a verb rather than as a noun' (2014: 76 in Doel, 2000:125): it is the activity created by the coming together of people and places, readers, writers and characters, texts and plot events. Narrative space, then, is something complex and contingent that arises through various spatial practices, making it, perhaps, more appropriate to speak not of narrative space but of a narrative's multiple spatialities. Where Hones attends to how these spatialities are produced within the text, this chapter indicates that what happens within the narrative world is tied to things that happen or stretch beyond its material confines and imagined world. *Fraternity's* route through the world, then, was neither a straightforward nor singular one of inspiration, followed by creation and reception. Instead, these positions or moments were much more uneven and the relationships that are sketched here give some indication of this, being an example of what we might term reception-in-creation: the worlds of the author and their invited reader intersect in an effort to define and refine the epistemological and ontological integrity of the narrative world.

The letters at the heart of this chapter not only reveal that *Fraternity* travelled differently through the world, but also that on its return to Galsworthy the world of the novel had been stretched, perhaps even distorted, in such a way that it needed to be made more resilient before it travelled further. Galsworthy had been left with two almost opposing readings; where Garnett sought to quieten Galsworthy's authorial voice and fashion a narrative world that had creative opportunities for the reader, Conrad wanted to close down opportunities for the reader's imagination by instigating a more powerful authorial narration. Where each had brought their own conception of Galsworthy and his writing style to bear on their reading of the novel and their perception of the world it should produce on general reception, in revising *Fraternity* Galsworthy had to decide which version of himself was to go forward. This decision coalesced most tellingly, Garnett observed, at the novel's conclusion:

> I once spent forty-eight hours, during a visit to Wingstone, trying to persuade him that it was impossible for him to make Hilary, the hero of Fraternity, elope with the little model. He confronted me with a letter in which Conrad, either out of devilment or in a fit of irritation, had advised him to the contrary, and it was touch-and-go for days as to which of his literary advisors would win!
> (Garnett 1934: 13)

In Garnett's mind the elopement of Hilary and Ivy (the little model) sat awkwardly with the novel's very progressive, artistic impulse. It was something of a clumsy ending that closed down the narrative world in a manner too decisive for Galsworthy's new found style. Conrad, however, in his belief that Galsworthy should remain loyal to his realist roots saw the elopement as the didactic device for encouraging greater intercourse between social classes, thereby drawing the narrative world to a natural and expected closure. As we follow Garnett and Conrad as they read themselves into the practice of *Fraternity* we see how reception-in-creation contests and reworks the social world and social relationships that underpin the novel's development. In the end, the generative effect was the making of a new narrative world for Galsworthy, for as he wrote to Garnett, 'after much thought I've decided to shape the ending to your thinking – going back to my very first conception – I think I can do it and that it will be more in accord with the conception of Hilary and the girl as drawn' (Garnett 1934: 167).

Opening the envelopes and prying into the correspondence that darted between writers and their literary confidents is one way of elucidating, more precisely and substantively, the way real and imagined worlds come together, inform and change one another. The world-in-the-letter is an unstable agglomeration: it is an interpretation of one world and the conveyor of another, and as it travels between and is mulled over within social worlds we begin to see something of its narrative consequences: it sparks discussion and disquiet, demands interpretation and invites responses and encourages refinements, reformulations and re-writings. In short, it is creative: it does things and exploring this creative force allows us to see how others are involved in the production of narrative space across the event of the text. Once a text is in the world readers bring many different worlds, real and imagined, to bear on the narrative and consequently on the production of space therein, but readers are no less productive before the text. Here though, the consequences of their reading can have very profound material effects on the very construction of the narrative world.

Note

1 See Chapter 2 for a greater discussion of *The English Review* and the relationship between Ford, Conrad and Galsworthy.

References

Altman, J.G. 1982. *Epistolarity: Approaches to a Form*. Columbus: Ohio State University Press.

Anon. 1906. The Man of Property. *The Times Literary Supplement*, (30 March 1906).

Anon. 1895. Almayer's Folly. *The Spectator*, (19 October 1895).

Barker, D. 1963. *John Galsworthy: The Man of Principle*. London: Heinemann.

Barnett, C. 1996. A choice of nightmares: narration and desire in Heart of Darkness. *Gender, Place and Culture*, 3 (3), pp.277–291.

Barton, D. and Hall, N. 2000. Introduction. In D. Barton and N. Hall, eds. *Letter Writing as a Social Practice*. Amsterdam: John Benjamins, pp.1–14.

Bradbury, N. 2001. Dickens and the form of the novel. In J.O. Jordan, ed. *The Cambridge Companion to Charles Dickens*. Cambridge: Cambridge University Press, pp.152–166.

Bushell, S. 2009. *Text as Process: Creative Composition in Wordsworth, Tennyson, and Dickinson*. London: University of Virginia Press.

Clay, C. 2006. *British Women Writers, 1914–1945: Professional Work and Friendship*. Aldershot: Ashgate.

Conrad, J. 1990. *Joseph Conrad: Interviews and Recollections*. Edited by M. Ray, Iowa City: University of Iowa Press.

Conrad, J. 1906. A middle class family. *Outlook*, (31 March 1906).

Curry, M. 1996. *The Work in the World: Geographical Practice and the Written Word*. Minneapolis: University of Minnesota Press.

Doel, M. 2000. Un-gluncking geography: spatial science after Dr Seuss and Gilles Deleuze. In M. Crang and N. Thrift, eds. *Thinking Space*. London: Routledge, pp.117–135.

Doyle, C. 1998. The writer tells: the creative process in the writing of literary fiction. *Creativity Research Journal*, 11 (1), pp.29–37.

Dupré, C. 1976. *John Galsworthy*. London: Collins.

Galsworthy, J. [1909] 1995. *Fraternity*. New York: Carroll and Graf.

Galsworthy, J. [1906] 2001. *The Man of Property*. London: Penguin.

Garnett, E. ed. 1934. *Letters from John Galsworthy, 1900–1932*. London: Jonathan Cape.

Garnett, E. 1904. *Letter from Edward Garnett to John Galsworthy*. No.17 12 June 1904. The Edward Garnett Papers. Eton: Eton College Library.

Gindin, J. 1987. *John Galsworthy's Life and Art: An Alien's Fortress*. Ann Arbor: University of Michigan Press.

Hayward, J. 1997. *Consuming Pleasures: Active Audiences and Serial Fictions from Dickens to Soap Operas*. Kentucky: University of Kentucky Press.

Hegglund, J. 2012. *World Views: Metageographies of Modernist Fiction*. New York: Oxford University Press.

Hones, S. 2014. *Literary Geographies*. London: Palgrave Macmillan.

Hones, S. 2008. Text as it happens: literary geography. *Geography Compass*, 2 (5), pp.1301–1317.

Iser, W. 1993. *Prospecting: From Reader Response to Literary Anthropology*. Baltimore: John Hopkins University Press.

Karl, F. and Davies, L. eds. 1991. *The Collected Letters of Joseph Conrad: Vol.4, 1908–1911*. Cambridge: Cambridge University Press.

Kauffman, L. 1986. *Discourses of Desire: Gender, Genre, and Epistolary Fictions*. New York: Cornell University Press.

Kitchin, R. and Kneale, J. 2001. Science fiction or future fact? Exploring imaginative geographies of the new millennium. *Progress in Human Geography*, 25 (1), pp.19–35.

LeFevre, K.B. 1987. *Invention as a Social Act*. Carbondale: Southern Illinois University Press.
Marrot, H.V. 1936. *The Life and Letters of John Galsworthy*. New York: Charles Scribner's Sons.
Mucignat, R. 2013. *Realism and Space in the Novel, 1795–1869*. Farnham: Ashgate.
Peters, J.G. 2001. *Conrad and Impressionism*. Cambridge: Cambridge University Press.
Salska, A. 1998. Dickinson's letters. In G. Grabher, R. Hagenbuchle and C. Miller, eds. *The Emily Dickinson Handbook*. Amherst: University of Massachusetts Press, pp.163–180.
Saunders, A. 2010. Literary geography: reforging the connections. *Progress in Human Geography*, 34 (4), pp.436–452.
Shapin, S. 1994. *A Social History of Truth: Civility and Science in Seventeenth-Century England*. London: University of Chicago Press.
Shotter, J. 1993. *Conversational Realities: Constructing Life through Language*, London: Sage.
Socarides, A. 2012. *Dickinson Unbound: Paper, Process, Poetics*. Oxford: Oxford University Press.
Stanley, L. 2004. The epistolarium: on theorizing letters and correspondence. *Auto/Biography*, 12 (3), pp.201–235.
Stanley, L. Salter, A. and Dampier, H. 2012. The epistolary pact: letterness and the Schreiner epistolarium. *Auto/Biography*, 27 (2), pp.263–293.
Whyman, S. 2007. Letter writing and the rise of the novel: the epistolary literacy of Jane Johnson and Samuel Richardson. *Huntington Library Quarterly*, 70 (4), pp.577–606.
Withers, C. 2004. Writing in geography's history: Caledonia, networks of correspondence and geographical knowledge in the late enlightenment. *Scottish Geographical Journal*, 120 (1–2), pp.33–45.

Epilogue

Travelling objects

This book began with Trollope, it closes with Bennett and with a last peek into one of his interiors and a consideration of its significance to his literary practice. This time it is not a private interior we enter but a very public one: the Establishment Duval, Rue de Clichy, Paris. The Duval restaurants, or Bouillons Duval as they were known in France, were similar in some ways to the Lyons Corner Houses of Britain. They were, as a Baedeker Guide to Paris observed:

> restaurants á la carte of a cheaper kind, managed in a peculiar way. The food is generally good but the portions are rather small, and the cost of a meal can scarcely be less than 2 ½-3 fr. (25 c. extra for a napkin). The guests are waited on by women. These houses are very popular with the middle and even upper classes, and may without hesitation be visited by ladies. Each guest on entering is furnished with a card (*fiche*), on which the account is afterwards marked. A fee . . . is left on the table for attendance; the bill is then paid, either at the desk or through the waitress.
>
> (Baedeker 1907: 17)

The Establishment Duval on the Rue de Clichy was one of Bennett's favourite dining spots in Paris. It was his 'usual restaurant' where he was 'expected, and where my usual waitress advised me in the selection of my lunch' (Bennett 8 November 1903, 1932: 125). Whether Bennett liked the restaurant's cheapness, the service, the fact that he was known, its convenience or the efficiency of its set-up is difficult to know, but that it was an important site in his early Parisian life is evident in the postcard he sent, to his mother, of the restaurant's interior.

Postcards are, what Veronica della Dora (2009) terms, travelling objects; they are portable, material things that are simultaneously representational and performative. As well as being visual texts to be read, they are active entities that interact with thinking, feeling bodies. 'Reading' this postcard as a representational form we can say something about the interior Bennett inhabited quite regularly. It was an interior in keeping with Bennett's love of the French Empire style. This style, as earlier discussed, used dark wood, often mahogany. Furniture was often spare, but massive in style with a tendency towards symmetrical designs with

Figure E.1 Interior image of the Duval Establishment, Paris.

Image courtesy of The Potteries Museum and Art Gallery, Stoke-on-Trent, Arnold Bennett papers (STKMG 1953.SH.20).

strong geometric themes. Décor, as can be seen from the room's panelling, was rich in colour, with a propensity towards ormolu mounts and gilt framing. With high ceilings and tables and chairs stretching back into the distance, the postcard gives the impression of a large, airy and busy restaurant. The uniform rows of tables also suggest a certain utilitarianism. This is not an exclusive eating place, but rather, one that fulfils a middle-class need for affordable quality. The relative openness of the restaurant also tells us something about its sociality, drawing us to consider the performances that lie behind the form. Unlike the cloistered booths of more exclusive establishments, this was a restaurant that lent itself to interaction. The wide thoroughfares between tables speak of human traffic, while the restaurant's high ceilings and expansiveness suggest that at busy times there was an audible hubbub in this place. This was not a place for quiet contemplation, for one could never predict who one would encounter and how:

> Last night, when I went into Duval for dinner, a middle-aged woman, inordinately stout and with pendent cheeks, had taken the seat opposite to my prescriptive seat. I hesitated as there were plenty of empty places, but my

> waitress requested me to take my usual chair. I did so, and immediately thought: 'With *that* thing opposite to me my dinner will be spoilt!'
>
> (Bennett 18 November 1903, 1932: 130, emphasis in original)

Bennett is a regular, he is known and has his place within this social world. His place is so certain that if you look carefully at the image you will see he has marked it with a cross, observing on the postcard's reverse: 'Here you have a postcard of the restaurant where I usually take my lunch. It does not exaggerate the size of it, which is enormous. I always have the same table; it is on the left. I have marked it with a cross. I am working . . . on my new humorous book' (Bennett 1903). From this position we know that Bennett sat adjacent to the aisle, looking into the restaurant's interior and was able to observe, but also likely to be able to feel, the jostle of its comings and goings.

Leaving this interior for the moment, let us turn to its representational form: the postcard, which at the turn of the twentieth century was transforming communication. In 1902 the Post Office permitted one whole side of a postcard to be taken up with an illustration, dividing the other side in two with one section for the address and the other for a short message. Other European countries adopted similar regulations, unleashing a textual form that lent itself to brevity, a more informal style of writing and the opportunity to rely on the image to stand in for explanatory prose (Gillen 2013). Unlike letter writing that tended to require the giving up of time and seclusion within a quiet, private space, a postcard could be dashed off quickly, written as easily at the table amidst the debris of dinner as it could on one's knee in a cramped train carriage. The limited space of the postcard combined with its portability as an inscribable object meant that it fostered particular types of communication (Andriotis and Mavrič 2013). It was often a more *ad hoc* way of keeping in touch, a space for updating, a site for observing everyday life, a substitute for a daily hello, in short, a kind of social shorthand. Thus, it is unsurprising that the Edwardian postcard has been seen as opening up a new epistolary space that chimed with the era's growing social and spatial mobility (Gillen and Hall 2010). Unlike the letter though, which played a crucial role in literary practice, the postcard's significance, given the style afforded by the form, seems less obvious. Nor, if we return to the interior the form depicts is it a place that seems particularly conducive to creative writing. Why then, conclude with this postcard?

Place and scene in the Establishment Duval

The answer lies in what we have come to know about creative writing. Writing both occupies and exceeds place; it happens in locations, but this happening is just one moment or aspect of a literary practice that stretches backwards and forwards, turns and returns, is iterative and reiterative. Thus, it has been suggested that a fuller understanding of the place of writing arises where we read it productively with the idea of the scene: the sense of the on-goingness or the *longue-durée* of creation. On the one hand, this captures the livedness and the discontinuities of literary practice: the way life frustrates, hinders and helps progress. On the other hand, it draws our attention to the relationality between practices. Writing is not a stand-alone happening, it springs from somewhere and something. We might variously call this inspiration, motivation or impetus, but remove these labels and at a more mundane level it is bound up with how we dwell, move and socialise within the world, and the scene allows us to appreciate how these practices nest inside and give shape and substance to writing. The scene then, enables us to move beyond literary practice as purely an act of writing that occurs within a cloistered garret, and comprehend it as a social happening that expands, contracts and transforms as different practices, people and things fall within and give shape to it.

The Establishment Duval is symptomatic of this more open and on-going sense of place and the distributed creativity (Hawkins 2013) or practice it acknowledges. It was large and through its doors came all manner of people, whose comings and goings and general chatter would have mixed with the sounds and rhythms of a restaurant designed to be busy. This may well have not been a conducive place to write, but its happenings were catalytic to the span of Bennett's literary practice as we discover if we turn to his novel, *The Old Wives' Tale*, which was published in 1908. Written five years after Bennett first went to live in Paris, it narrates the lives of two sisters, one who stays in the Potteries and one who escapes to Paris. In the preface to this novel Bennett returns the reader to the November evening he spent at the Establishment Duval in 1903. In this version of events the lady is not sat at his table, but merely comes into the restaurant to dine:

> She was fat, shapeless, ugly, and grotesque. She had a ridiculous voice, and ridiculous gestures. It was easy to see that she lived alone, and that in the long lapse of years she had developed the kind of peculiarity which induces guffaws among the thoughtless. She was burdened with a lot of small parcels, which she kept dropping. She chose one seat; and then, not liking it, chose another; and then another. In a few moments she had the whole restaurant laughing at her . . . I reflected, concerning the grotesque diner: 'The woman was once young, slim, perhaps beautiful; certainly free from these ridiculous mannerisms. Very probably she is unconscious of her singularities. Her case is a tragedy. One ought to be able to make a heart-rending novel out of the history of a woman such as she' . . . It was at this instant that I was visited by the idea of writing the book which would ultimately become The Old Wives' Tale.
>
> (Bennett [1908] 2007: 31–32)

From a seemingly small incident came the beginnings of what would prove to be one of Bennett's most popular novels. It was an incident that occurred in one place, but which travelled with Bennett, materially in his journal, but also mentally as he frequently returned to it but each time put it aside to 'write novels of smaller scope' ([1908] 2007: 33). That Bennett recorded this incident in the first place and obviously returned to his journal to remind himself of events and happenings discloses another line or trajectory of his literary practice. It is a line that leads back to the de Goncourt brothers whose diaries of everyday Parisian life Bennett took as a model for his own. The brothers framed their diary as a day-to-day autobiography of a place, recording all manner of incidental experiences in order to document the everydayness of life (de Goncourt and de Goncourt 2007). It is a diary in which we find the likes of Émile Zola, Guy de Maupassant, Gustav Flaubert and Honoré de Balzac. These were all writers Bennett admired, principally for their ability to find interest in what appeared as the dullest and drabbest aspects of life. Moreover, they were all tied to Montmartre, a district associated with Paris's avant-garde art scene, and on whose cuff we find the Rue de Clichy and the Establishment Duval. Coincidences perhaps, but as with John Galsworthy and the neighbourhood of Addison Road, places are repositories of lives, practices and cultural capital that become nexuses for haphazard encounters, the transfer of ideas and crucibles for reinventing and reinforcing the making of something new.

If the location of the Establishment Duval was a wormhole into Paris's literary history, its materiality as a travelling, inscribable object expresses something even more expansive about the scene of writing. As an act of ordinary writing, this postcard disrupts the one-dimensional notion of the novelist as only a novelist who we encounter only in and through their professional world. By disclosing the more prosaic writings and everyday relationships through which a writer lived their life, this postcard punctuates the hermetic world of the literary elite. It acknowledges both the *others* who may haunt the wings of this world and the way self-identity fractures and reforms across it. If we take these others, they are neither critics nor critical friends, they may be provincial not metropolitan and wholly uninterested in the whirligig of literary life, but they signify that the geography of the writer's world is more than their immediate surroundings or professional networks. Similarly, the postcard performs another version of Bennett. This is not an author but a son writing home. In conveying and recounting where he usually had lunch, in stressing the restaurant's size and by placing himself very firmly in the picture so that his positionality travelled across time and space, the postcard created and shared with those not present a vicarious experience of everyday life (Price 2013). Thus, Bennett's postcard to his mother was a portal through which she could come to know and imagine something of the everyday world, with its occupations and cares, her son was living. Or, put another way, it performed a version of Bennett and of his everyday world in a place (and to a person) that was no longer a directly experienced part of his everyday world. The postcard was a travelling object that was expressive of the ordinariness of the writer's life: the gamut of things that were not creative writing, that were predictable, routine and unglamorous, in short that which probably constituted the bulk of lived experience. This is, then, another way into the *longue-durée* of writing, and in particular to the way creative writing is discontinuous and jerky, something rough not smooth, meandering not linear. Moreover, it folds us back into the ongoingness, or temporality, of the narrative world, by returning us to the idea of the filler: to those aspects of life that form the substance of most imagined worlds (Moretti 2006). Most lives, if we had to characterise them, would be composed primarily of fillers not events, everyday things easily forgotten and effaced by more significant but rarer happenings. In the imagined world we remember the events, but it is ordinary things that connect these discrete happenings into a coherent and ongoing scene of action. We think little about them, they are perhaps a translucent fabric veiling the whole, but without them there would be no imagined world. As a travelling object the postcard reminds us both of the way near and far are never as distant or as separate as they seem, but also of the way the real and the imagined come together, not as copies of one another but as entities that jostle, mingle and leave traces of themselves on the other.

Travelling across thresholds

Those familiar with British television of the early 1990s may well recall *Through the Keyhole*. Hosted by Lloyd Grossman it offered viewers glimpses of celebrity homes, asking them to sleuth, through carefully laid clues, to whom the house belonged. Various keyholes have been violated here; possession has been taken of the keys to several of Bennett's homes, sight has been had of the work plans Trollope so assiduously kept and kept with him and the private letters that passed between Galsworthy and his friends have been steamed opened and their contents laid bare. Of course, the analogy only takes us so far, and is perhaps a little off-beat and frivolous, but it is also a catalyst reinforcing that we can only ever catch glimpses of past literary practices. These glimpses come to us in various forms. Some begin in the lived world, as chinks and fissures in the biographies we have of this world that come to the fore. Others begin in the text itself, alighting on those wormholes where the contours of the lived world are visible or the fabric of the imagined world appears thin, enabling us to see, and to follow, the pathways between the imagined world and the real world. Then there are those that emerge in the ephemera of life, in the textual traces, things and objects that are the stuff of life, but which we frequently overlook as inconsequential to what we understand as literary practice. Although only glimpses, they are glimpses of a world of relationships not individuals, of people not just authors, and of process not stasis. It is a social world that is always ongoing.

As we enter the worlds that lie beyond the keyholes we also enter worlds that have, for a long time, been locked away and hidden from view within literary geography. The worlds before the text are like the rooms of a Victorian mansion that are sheathed in dust sheets and closed up while the social season takes place elsewhere. The only reason to enter these rooms was to uncover simple meaning about textual origins. Thus, they have been left alone to gather dust, while literary geography, like literary studies, has lived largely, although not exclusively, within the text. Behind these doors though, is a complex, lively and discontinuous world, and the more doors we open the more glimpses we get of this rich world. Nicholas Bourriaud (2002) once observed of visual art that the art work was only ever a dot in a line; there was a world after it, but also one before it and to fully understand it we needed to be familiar with all the worlds that stretched from it. Although we should repudiate the line analogy, for if anything should be clear by now it is that the life or lives that come before the art work, or in this instance the novel, are multiple, erratic and unstable, it is helpful in forcing us to think of the text as only one moment within a web of geographies that stretch backwards, forwards and sideways. At times, these trajectories may stall, appearing to coalesce or pause in place, but from these spatial fixes new lines and lives steal out, both complicating and enriching the world in which the text is made.

References

Andriotis, K. and Mavrič, M. 2013. Postcard mobilities: going beyond image and text. *Annals of Tourism Research*, 40, pp.18–39.

Baedeker, K. 1907. *Paris and its Environs, with Routes from London to Paris: Handbook for Travellers*. London: Fisher Unwin.

Bennett, A. 1932. *The Journals of Arnold Bennett, Vol. I: 1896–1910*. N. Flower, ed. London: Cassel & Co.

Bennett, A. [1908] 2007. *The Old Wives' Tale*. London: Penguin.

Bennett, A. 1903. *Postcard from Arnold Bennett to his Mother*. STKMG.1953.SH.20. Stoke: Potteries Museum and Art Gallery.

Bourriaud, N. 2002. *Relational Aesthetics*. Dijon: Du press du reel.

de Goncourt, J. and de Goncourt, E. 2007. *Pages from the Goncourt Journal*. G. Dyer, ed. New York: New York Review Books. Trans. R. Baldick.

della Dora, V. 2009. Travelling landscape-objects. *Progress in Human Geography*, 33 (3), pp.334–354.

Gillen, J. 2013. Writing Edwardian postcards. *Journal of Sociolinguistics*, 17 (4), pp.488–521.

Gillen, J. and Hall, N. 2010. Edwardian postcards: illuminating ordinary writing. In D. Barton and U. Papen, eds. *The Anthropology of Writing: Understanding Textually Mediated Worlds*. London: Continuum.

Hawkins, H. 2013. *For Creative Geographies: Geography, Visual Arts and the Making of Worlds*. London: Routledge.

Moretti, F. 2006. Serious century. In F. Moretti, ed. *The Novel: History, Geography, and Culture, Vol.I.* Oxford: Princeton University Press, pp.364–400.

Price, C. 2013. Tokens of renewal: the picture postcards as a secular relic of re-creation and recreation. *Culture and Religion: An Interdisciplinary Journal*, 14 (1), pp.111–130.

Index

Page numbers appearing in italics refer to figures.

For Product Safety Concerns and Information please contact our EU
representative GPSR@taylorandfrancis.com
Taylor & Francis Verlag GmbH, Kaufingerstraße 24, 80331 München, Germany

www.ingramcontent.com/pod-product-compliance
Ingram Content Group UK Ltd.
Pitfield, Milton Keynes, MK11 3LW, UK
UKHW022124190926
475702UK00013B/1625
* 9 7 8 0 3 6 7 2 4 4 1 4 9 *